INTERNATIONAL MINING FORUM 2005

PROCEEDINGS OF THE 6TH INTERNATIONAL MINING FORUM 2005
23-27 FEBRUARY 2005, CRACOW – SZCZYRK – WIELICZKA, POLAND

Underground Mining:
New Technologies,
Safety and Sustainable Development
International Mining Forum 2005

Edited by

Eugeniusz J. SOBCZYK
Polish Academy of Sciences, Mineral and Energy Economy Research Institute, Cracow, Poland

Jerzy KICKI
*AGH – University of Science and Technology, Department of Underground Mining,
Cracow, Poland
Polish Academy of Sciences, Mineral and Energy Economy Research Institute, Cracow, Poland*

Taylor & Francis
Taylor & Francis Group

LONDON AND NEW YORK

Published by: Taylor & Francis
2 Park Square, Milton Park, Abingdon, Oxon, OX14 4RN
270 Madison Ave, New York NY 10016

Transferred to Digital Printing 2007

ISBN 0415 375525

Publisher's Note
The publisher has gone to great lengths to ensure the quality of this reprint but points out that some imperfections in the original may be apparent

International Mining Forum 2005, Sobczyk & Kicki (eds) © 2005 Taylor & Francis Group, London, ISBN 0415 375525

Table of Contents

Preface

"Development that meets the needs of the present generation without undermining the capacity of future generations to meet their needs". That is how, for the first time, *sustainable development* was defined by Brundtland Commission. Sustainable development is based on social, economical and environmental pillars. Each of these pillars is addressed in the speeches given at the International Mining Forum.

Having been updated with new technological solutions, an interesting picture is given of the experiences of different countries in the field of sustainable development. Without a doubt, those countries that first put it into practice, have the most experience in applying such progress. Such trial is, I think, important in understanding the subsequent results.

Mining is a very specific activity with strong social impacts, and a radical influence on the environment. It is therefore very sensitive to the ideals of sustainable development. The up-to-date effects of implementation have not been as successful as one would have hoped. It is our vision, that thanks to forums such as this, queries can be addressed, ecosystems improved and new inter-personal relations formed and strengthened.

This work comprises technical papers that were presented at the International Mining Forum in Kraków – Szczyrk – Wieliczka, Poland, held on 23–27 February 2005.

The major themes of IMF 2005 were:

- New technology in underground mining.
- Safety on mines.
- Mining under conditions of sustainable development.

Numerous fields of the world's mining industry prove that besides sustainable development being possible, there are also new developments within this sector. Needless to mention this is due to consumers' demand for energy and resources.

This book is addressed to researchers and professionals who work in the fields of underground mining technology, rock engineering or mine management.

The topics discussed in this book are:

1. Royalties in European Community Countries.
2. The problems of mine closure.
3. Rock engineering problems in underground mines.
4. Trends in the mining industry.
5. New solutions and tendencies in underground mining technology.
6. The impact of mining technology on the environment.
7. The chosen GIS application.
8. Some methods of mineral projects risk analysis.
9. Current problems of Chinese coal mines.

The International Mining Forum was held thanks to the support of the Chair of Underground Mining, the Faculty of Mining and Rock Engineering of the University of Science and Technology (AGH), Mineral and Energy Economy Research Institute of Polish Academy of Science in Cracow, KGHM Polska Miedz SA, Jastrzebska Coal Company, Katowice Coal Holding, CUPRUM Ltd, MIDO Ltd, MMDE ZOK Ltd., ELGÓR+HANSEN SA. The organizers would also like to express their gratitude to all the other individuals, companies and institutions, who helped in bringing the Forum into being. We hope that the Forum will throw new light on mining and will encourage the exchange of interesting experiences!

Jerzy Kicki
Chairman of the Organizing Committee 2005

Organization

Organizing Committee:
Jerzy Kicki (Chairman)
Eugeniusz J. Sobczyk (Secretary General)
Artur Dyczko
Jacek Jarosz
Piotr Saługa
Krzysztof Stachurski

Advisory Group:
Prof. Volodymyr I. Bondarenko (National Mining University, Ukraine)
Mr. Wojciech Bradecki (State Mining Authority, Poland) – Chairman of IMF 2005
Prof. Jan Butra (CUPRUM Ltd., Poland)
Dr. Alfonso R. Carvajal (Universidad de La Serena, Chile)
Prof. Piotr Czaja (AGH – University of Science and Technology, Poland)
Prof. Bernard Drzęźla (Silesian University of Technology, Poland)
Prof. Józef Dubiński (Central Mining Institute, Poland)
Prof. Jaroslav Dvořáček (Technical University VSB, Czech Republic)
Prof. Paweł Krzystolik (Experimental Mine Barbara, Poland)
Prof. Garry G. Litvinsky (Donbass State Technical University, Ukraine)
Prof. Eugeniusz Mokrzycki (Polish Academy of Sciences, MEERI, Poland)
Prof. Roman Ney (Polish Academy of Sciences, MEERI, Poland)
Prof. Jacek Paraszczak (University of Laval, Canada)
Prof. Janusz Roszkowski (AGH – University of Science and Technology, Poland)
Prof. Stanisław Speczik, (Polish Geological Institute, Poland)
Prof. Anton Sroka (Technical University – Bergakademie, Germany)
Prof. Mladen Stjepanovic (University of Belgrade, Yugoslavia)
Prof. Antoni Tajduś (AGH – University of Science and Technology, Poland)
Prof. Kot F. v. Unrug (University of Kentucky, USA)
Dr. Leszek Wojno (Australia)

International Mining Forum 2005, Sobczyk & Kicki (eds) © 2005 Taylor & Francis Group, London, ISBN 0415 375525

Royalties as the Tool of Raw Material Policy of the European Union Countries

Jaroslav Dvořáček
VŠB – Technical University of Ostrava, Faculty of Mining and Geology. Ostrava, Czech Republic

ABSTRACT: The sustainable development is connected with intensity of raw materials exploitation. This intensity depends on raw material policy of particular state that is influenced by conditions of granted mining rights and licences including the mining royalties. The contribution deals with the problem of royalties in "old" and "new" European Union countries as a part of their raw material policy.

KEYWORDS: Raw material policy, mining legislation, royalties

1. INTRODUCTION

In principle, the sustainable development in the sphere of raw materials is connected to the rate of exhausting of natural resources, so with the mineral raw materials mining on one hand but also with utilizing of recycled materials on the other. Both these activities appertain to raw material policy of the state having the effect on business activity within the mining industry. This raw material policy can have the form of an individual document, however, it must be often deduced from various legal regulations, government activity or declarations of key governmental representatives (Otto 2002). The legislative base as well as practical activity of the ministries and the governments should take into account the specific features of the business activity in the mining industry. Among the most important ones is the close relation between the parameters of a mineral raw materials deposit and the economic results, duration of a mining activity as well as environmental impacts. Other characteristic features are represented by the irreversible character of mining activity and, in most cases, a large number of people involved in the process as owners of the mining companies, the land on which mineral deposits are found and owners of the deposit itself.

The multitude of owners causes that the prospecting and mining activities exclude certain lots of land from their original determination causing herewith losses of production and leading to the rise of damages. Despite the fact that the goal of reclamation is to return the territory back to its original state as far as possible, it can't be done in all cases. The mining activity itself decreases the minerals resources, decreasing herewith the value of the property of the deposit's owner.

So, it can be stated that the prospecting and mining activities lead to the decrease of property value and damages. It is not surprising that the owners of land and deposits of mineral raw materials require certain compensation. This compensation for damages and property devaluation, in most cases, has the following forms: (i) settlements from the prospected territories, (ii) settlements from mining areas, (iii) settlements (royalties) from the exploited mineral.

2. COMPENSATION FOR DAMAGES AND PROPERTY DEVALUATION

The base of compensation for damages and property devaluation due to prospecting and mining activities is represented by the problems of ownership of land with the mineral raw materials deposits but first of all by the problems of mineral raw materials deposits ownership.

Generally, it can be stated that the settlements from territories being prospected and settlements from mining areas are represented by payments by prospecting or mining company to the owners of relevant lots of land for allowing the prospecting or mining activity to take place, by which also the damages arisen due to other utilization of lots of land are compensated. The royalties represent the form of compensation to the mineral raw materials deposit owner for conveying the right to appropriate the mineral exploited and, herewith, to decrease the property value of the mineral raw materials owner.

The question of mineral ownership becomes significant in the moment of the deposit's discovery. If the mining company wants to sell the exploited minerals it must become their owner or obtain the right to mine them.

In some countries the minerals belong to the owner of the land, then this ownership assignment can be carried out in situ or after their exploitation. In most cases, the mineral resources belong to the state. Granting a mining licence does not transfer the minerals ownership but only provides the right to mine. The state declares herewith its interest to retain consistent sovereignty over its mineral resources.

If we confine ourselves to the European Union countries the problems of mineral raw materials deposits ownership can be characterized as follows.

I. The owner is the state amd the ownership contains:
a) all mineral resources without exceptions: Lithuania, Hungary, Slovenia,
b) all mineral resources if not set otherwise: Netherlands (the ownership of minerals exploited on the basis of a licence is transferred on the licence holder), Estonia (the ownership of deposit belongs to the land owner if he exploits it for his own needs),
c) the kinds of mineral raw materials specified in a determined way:
   - the reserved minerals word for word or in translation equivalent: France, Ireland, Czech Republic, Slovak Republic (all the reserved minerals), German Federal Republic (free mineral resources), Austria (free as to the mining), Sweden (licensed minerals), Cyprus (the materials exploited), Latvia (mineral resources of state importance), Poland (deposits which constitute important parts of real estates), Italy (minerals of the 1st category), Belgium (substances which can be exploited in an underground manner), Luxembourg (deposits below 6 meter depth), Spain (state reserves),
   - listed mineral raw materials: Greece (hydrocarbons, fuels, radioactive minerals, geothermal power, salt), Great Britain (coal, oil, gas, uranium, gold, silver), Denmark (underground deposits of oil, gas, salt), Malta (hydrocarbons, minerals in continental shelf), Portugal (ore deposits, hydro-mineral and geothermal resources). The minerals named in these countries are not called reserved ones compared with the previous categories where the term „reserved minerals" is, as a rule, supplemented by their enumeration already in mining industry legislation.
II. The owner is the owner of the land in case that the owner of deposit is not the state:
   - non-reserved minerals word for word or in translation equivalent: France, Ireland, Czech Republic, Slovak Republic (all the non-reserved minerals), Belgium (the substances which can be mined in an opencast manner), Luxembourg (deposits up to 6 meters depth), Italy (minerals of the 2nd category), Denmark (shallow deposits), Sweden (besides the licensed minerals), Latvia (deposits outside the state importance), Great Britain (other mining industry's authorizations), Portugal (minerals not included under the ore deposit category),
   - deposits named expressly as an integral part of land: German Federal Republic, Austria, Poland, Cyprus (if the land owner has a license for their exploitation).

III. Deposit's finder: the specific feature is the so called *Doctrine of Mineral Claims* applied in Finland when the minerals deposit's finder has the priority to them regardless of who owns the land.

The demarcation of the lots of land ownership where the prospecting or mining activities are running and determination of ownership of the minerals deposit which will be or is already exploited is important from the viewpoint of directing of payments compensating the damages or property devaluation.

The situation is relatively simple in case of settlements from territories being prospected or settlements from mining areas. If these settlements are set legislatively they have the form of a set amount of money per unit area. The amount is determined legislatively or it is the matter of negotiation of mining company with the relevant land owner.

The more complicated is the situation of settlements for the minerals exploited, the so called royalties. These settlements are used often by governments because (i) they are related to each year production, they are certain and foreseeable (ii) they ensure the stable income of the state during a mine service life irrespective of the mining company's profit (iii) they are relatively simple to calculate, select as well as monitor (Otto and Cordes 2002).

The royalties for the minerals exploited can be related to:
- production volume or production value (to sales for a mine production),
- net yield from metallurgical plant, where the payment is set as a certain percentage of the difference between the income from the sales of production and costs of transport, melting, refining and marketing,
- the net profit where the payment is set as a certain percent from net profit gained by the mining company.

The simplest way of royalties determination is to calculate them from the natural output volume, especially in case of homogeneous production - e.g. coal, gravel sand, decoration rocks etc. However, a problem of production quality can arise here (power and coking coal, sand for building industry and glass industry), problem of distinction between the raw material subject to payment and waste rock, problem of dependence on reports by the mining company.

It is more suitable, from this point of view, to deduce the royalties from the production value which can be ascertained more accurately from accounting statements. The problem, however, is the setting of relevant market prices which can include the actual prices of home or world market, average prices, prices set mandatorily by the relevant authority etc. The fluctuation of prices causes the greater variability of the state income at using of production value compared with the physical indicators of output. The price used influences also the base in the royalties calculation deduced from the mining company's profit.

Generally, it is stated (Otto 1995) that the one best system for setting the royalties does not exist here and the system selected will depend on specific goals and administrative capability of the institution collecting the royalties for the minerals exploited.

In case of "old" as well as "new" European Union countries the following cases can be singled out in the sphere of royalties for the minerals exploited:
a) *the royalties are not expressly used in* Sweden and Greece,
b) *the royalties are settled on the basis of agreement with the land owner*: Finland,
c) *the royalties are settled as a percentage from production value determined in different way*: Belgium, German Federal Republic, Spain, France (hydrocarbons), Ireland (ores), Italy (hydrocarbons), Luxembourg, Netherlands (hydrocarbons), Austria (hydrocarbons), Great Britain (oil up to the year 2003), Czech Republic, Cyprus, Lithuania (hydrocarbons), Latvia (hydrocarbons), Hungary, Slovak Republic, Slovenia and Portugal,
d) *the royalties are settled as the fixed amount related to actual production volume*: France (payment to the territorial administration units in which the deposits are found), Ireland (industrial minerals), Great Britain (construction raw materials), Denmark, Estonia, Latvia (except hydrocarbons) and Poland.

3

Generally, it is can be stated that the comparison of quantitative amount of settlement between different countries is difficult because of differences in heterogeneity of natural resources and in calculation of the production value as well (price level, costs-deduction items). The economic support of mining of hydrocarbons from undersea deposits in the way of lower rates (Italy, Netherlands), decreasing of rates in newly exploited deposits (France, Lithuania) or full exemption of fees (Great Britain) can be traced up. In the case of Latvia the royalties have the form of licence for mining of planned volumes of mineral resources and exceeding the set output volume results in sanctions. In some countries the royalties are to motivate the use of recycled raw materials at the expense of the mining of natural resources – for example in the UK.

3. CONCLUSION

Generally, it can be stated that all European Union countries have regulations determining or effecting the raw material policy of the state. Most often they have the form of a nation-wide mining law, the exceptional ones are the mining laws issued by autonomous regions (Belgium). Some countries have the legislation for mineral raw materials including of or separately for hydrocarbons (Ireland, Austria, Portugal, Denmark, Malta, Portugal), the mining activity is influenced by the laws concerning the territorial planning (Great Britain) or law of a tax character (Latvia).

Generally, the rational, long-term utilization of natural resources is declared which, on one hand, ensures meeting the needs of national economy, on the other hand the environmental protection. The mining industry's legislation is supplemented by the environmental legislation.

If we evaluate the royalties from the viewpoint of mining company this is a question of additional costs. On the other hand, it is necessary to take into account that the right to exploit the deposit (belonging in many cases to other owner) is the base of mining industry's business activity because the mining company obtains the access to the most substantial component of its property. The royalties from the minerals exploited equalize, to a certain extent, the conditions of economic competition with the industries which must buy the subject of their activity from a producer or grow it – e.g. the processing industry.

Then, it can be stated that the royalties from the minerals exploited base on the non-renewable nature of raw material resources and their ownership, the main characteristic features of the mining industry.

REFERENCES

Otto J.M. 1995: Legal Approaches to Assessing Mineral Royalties. In: The Taxation of Mineral Enterprise. London, Graham & Trotman, 397 p., ISBN 85966105X.
Otto J.M. 2001: Fiscal Decentralization and Mining Taxation. The World Bank Group. Mining Department, March 2001.
Otto J.M., Cordes J. 2002: The Regulation of Mineral Enterprises: A Global Perspective on Economics, Law and Policy. Rocky Mountain Mineral Law Foundation. Westminster, Colorado 2002.
Parsons B. et al. 1998: Comparative Mining Tax Regimes. A Summary of Objectives, Types and Best Practices. Price Waterhouse Coopers.
Uberman R. et al. 2004: The Study of Minerals Planning Policies in Europe. AGH – University of Science and Technology, Department of Opencast Mining. Krakow, April 2004.
Information sources on Internet.

International Mining Forum 2005, Sobczyk & Kicki (eds) © 2005 Taylor & Francis Group, London, ISBN 0415 375525

Some Aspects of the Coal Industry Sustainable Development in China

Yuan Shujie

Anhui University of Science & Technology. Anhui Province, China

ABSTRACT: This paper presents the present situation of the coal industry in China, the issues of coal resources, work safety in coal mines and environmental protection, which influence the coal industry's sustainable development. The author gave some policy proposals for the Chinese coal industry sustainable development.

KEYWORDS: Coal industry, sustainable development

1. GENERAL SITUATION OF THE COAL INDUSTRY IN CHINA

China is the largest producer of coal in the world, the annual production reaches above 1,200 million tons, accounts for 1/3 of the total output of the world, the exported more than 30 million tons of coal annually account for 11% of coal trade in the world. China not only is the biggest coal producer in the world, but also the biggest coal consumer. In 2000 coal consumption in China accounted for 28% of the world total consumption. Annual coal consumption occupies 76% in the national total commercial energy consumption. Coal occupies about 70% of non-renewable energy production or consumption. Coal as the primary energy source for a quite long time won't change its status in the Chinese energy structure, but along with more and more strict environmental protection demands, the development and application of renewable energy, the proportion of coal in the energy structure will drop.

There are three types of coal mines in China: state-owned key mines, local state-owned mines (provincial government mines) and town and village owned mines. State-owned mines are typically large mines that have historically supplied local electricity and state owned industrial enterprises. Operation of these mines shifted in 1998 from the central government's Ministry of Coal to the provincial, regional or municipal governments. Town and village enterprises are generally small scale mines that often have no legal status and have poor health and safety standards. These mines generally operate with little or no safety equipment. Poor environmental health and safety performance of these mines and oversupply of coal in the domestic market prompted the central government to earmark 23,000 of these mines for closure in 1999. As of June 30. 2001, the Chinese government ordered that all illegal town and village small-scale coal mines are to close down and that all small scale enterprises that hold mining permits are to close down until they meet government safety standards. The programme saw 5,117 small-scale mines close in the year. By now in China there are 33,000 enterprises engaged in coal exploitation, including 571 state-owned key mines, 2000 local state-owned (provincial government) mines, about 31,000 town and village owned mines. Altogether approximately 6 million people work in coal mines (SACMSS 2002). In 2002 coal production structure adjustment initially saw effect, as the output of raw coal from state-owned key coal mines, local state-owned coal mines and village and town owned coal mines accounted for, respectively, 51%, 19% and 30% of the national total, as compared to 39%, 16% and 45% in 1996. The output from state-owned key coal mines was 711.6264 million tons, and compared to that in 2001 increased by 93.0515 million tons, or 15.04%; the output from local state-owned coal mines was 263.4524 million tons, an increase of 40.288 million tons, or 18.05%, comparing to 2001; the output from village and town owned coal mines was 418.2735 million tons, increased by 154.4188 million tons, 58.52%, comparing to 2001 (SACMSS Web 2004). In 2003 the coal market continuously got better, the high-speed economic development aroused great demand for energy and promoted large increase of coal output. The total coal output was 1.736 billion tons, comparing with that in 2002 increased by 343 million tons, i.e. grew 24.62%. Coal output from state-owned mines was 1.11 billion tons, accounting for 63.94% of the total coal output, dropped from 69.98% in 2002. Among the output from state-owned mines 830 million tons was from state-owned key mines, accounting for 47.81% of the total coal output, dropped from 51.07% in 2002; 290 million tons from local state-owned coal mines, accounting for 16.71% of the total output, dropped from 18.91% in 2002; 610 million tons from town and village owned coal mines, accounting for 35.16% of the total output, increased from 30.02% in 2002. Coal production from 2,001 to 2003 is shown in figure 1 (SACMSS Web 2004).

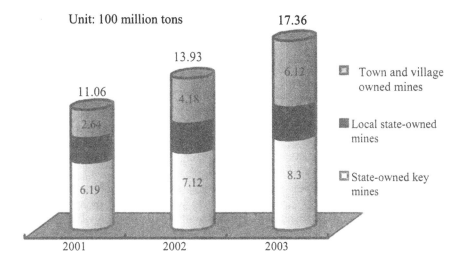

Figure 1. Coal production in China from 2001 to 2003

2. INFLUENCE OF RESOURCES ON COAL INDUSTRY SUSTAINABLE DEVELOPMENT

Coal resources in China are very abundant. Coal resources are distributed in extensive regions from the north to the south, demonstrating roughly three strips: from Tianshan Mountain to the south of Yinshan Mountain, including most parts of Northeast China, North China, Henan Province and the south part of East China, where various kinds of bituminous coal and anthracite occur, there are only small quantities of lignite; from Kunlunshan mountain through Qinling Mountain to the south of Dabieshan Mountain, including southwest, most parts of middle-south and southern areas of East China, there are mainly high rank coals, and a little amount of middle-rank bituminous coal and lignite. Verified resources occur mainly in five provinces: Shanxi, the Inner Mongolia, Shaanxi, Xinjiang and Guizhou. (Henan Coal Web 2004) The distribution of coal resources is shown in figure 2 (China Web 2004).

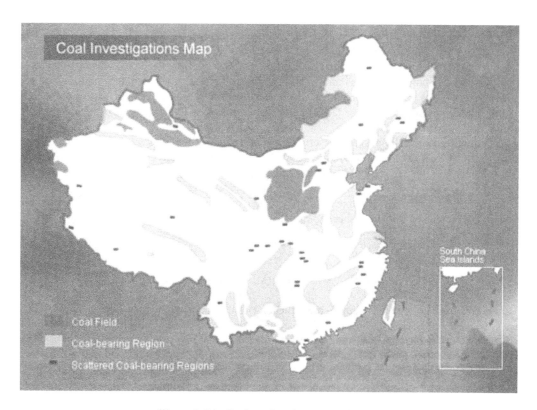

Figure 2. Distribution of coal resources in China

It's predicted, that to the depth of 2000 m below the surface the total amount of perspective resources of coal in China reaches 5,059.2 billion tons, only second to the Commonwealth of Independent States, occupying the second place in the world. By the end of 2000 the verified reserves amounted to 1,007.7 billion tons, with a resources verifying rate of 19.9%. According to the international general division methods, the economically available reserves of coal in China are 114.5 billion tons, approximately 11.6% of the world similar reserves (984.2 billion tons). According to coal output in 2000, 1 billion tons, verified reserves of coal in China guarantee 1000 years, economically available reserves guarantee 114 years of production. According to the present coal mine production and production discontinued or newly added in the future, it's estimated that in 2010 the coal production in China will be 1.21893 billion tons, in 2020 1.2531 billion tons, not including production of small coal mines with annual output below 30,000 tons and village and town owned coal mines. According to the above, the forecast coal output in 2010 will be 1.59 billion tons, in 2020 about 1.7 billion tons.

Because of obsolete technical equipments and mining methods, the recovery rate paces back and forth at a low level for a long time, especially large number of small coal mines and pits wantonly waste coal resources. According to statistics the average recovery rate in state-owned large-scale coal mines is about 40%, in village and town owned small coal mines only about 20%. In some small coal pits even 10 tons of resources are wasted for every 1-ton of extracted coal. But at present one half of coal output in China is produced by these wasteful small coal pits. It is estimated that in China the average annual coal output approximately reaches 1.15 billion tons, but actually as much as 4.3 billion tons coal resources must be consumed, which is nearly equal to the global coal output (SACMSS Web 2004).

Coal resources demonstrate such features as: huge reserves and complete variety but uneven distribution among different grades, with small reserves of high-quality coking coal and anthracite; wide distribution but a great disparity in abundance for different deposit locations, with large reserves in the western and northern regions and small reserves in the eastern and southern regions; a small number of opencast coal mines, most of which are lignite mines; and great variety of associated minerals found in coal seams. These factors cause some difficulties for the coal industry's sustainable development.

At present, economically available reserves account for only 30% of the verified coal reserves and most of them already have been mined out. Coal reserves are quite small. China has large population, the average coal resources possession per capita is approximately 234.4 tons, but the world average coal resources possession per capita is 312.7 tons, in the USA average coal resources possession per capita reaches as high as 1,045 tons, far higher than in China. For the coal industry sustainable development to satisfy demands of economic growth in China shortage of available coal resources will be a problem.

3. THE ASPECT OF WORK SAFETY IN COAL MINES

In China coal mine geological conditions are difficult. 95% of coal production comes from underground mines, where mining work is threatened by various natural hazards. Statistics show: in state-owned key coal mines gaseous mines account for 30%, more than 250 mines have hazard of coal and gas explosion. Since 1950 up to 1995 altogether 14,000 explosions of coal and gas occurred. 48% mines have hazard of spontaneous combustion, altogether 10,296 fires occurred in 32 years from 1953 to 1984. It equals to 1.48 of fire occurrence per one million tons of coal mined. The mines with coal dust explosion hazard account for above 90%. Since 1970 exploitation depth has increased by 130 m, along with exploitation depth increasing, high temperature and stress become problems (SCCCS 1995). Coal mine fires are one of the most serious natural hazards, which once caused huge economic losses and casualties. In the mine fire accidents exogenous fires account for less than 10% of the total occurrences, spontaneous combustion fires reach above 90% (Wang et al. 1994). One of the main causes for occurrence of fires in mines is spontaneous combustion of coal. About 51.3% of state-owned key coal mines have spontaneous combustion hazard, which is responsible for 90% of fire occurrences in coal mines. Only in 1999 in 87 large and middle scale coal mines, there were 315 accounts of fire areas sealing due to spontaneous combustion. It caused serious loss of coal resources and threat to miners' lives (Luo et al. 2003). In recent years in state-owned coal mines average annual occurrences of fire reached above 400 (Fushun Branch of... 2003).

According to the statistics of the State Administration of Work Safety, China produced 35% of the world's coal in 2003, but reported 80% of the total fatalities in coal mine accidents. Among accidents that occurred from January 2001 to October 2004, there were 188 with a death toll of more than 10, about one fatality every 7.4 days. The frequency of coal mine accidents in China is still very high. In 2003, the average coal miner in China produced 321 tons of coal a year; this is only 2.2 percent of that in the United States and 8.1 percent that of South Africa. The fatality rate for every million tons of coal output, however, is 100 times of that of the US and 30 times of the South Africa. Working conditions for China's coal miners need to be improved, about 600,000 miners to date are suffering from pneumoconiosis, a disease of the lungs caused by prolonged inhalation of dust. And the figure increases by 70,000 miners every year. The government has taken a lot of measures to improve work safety in coal mines. In 2000, China set up a national surveillance system to keep a close eye on the safety conditions of coal mines. In the following years, the government earmarked more than 4 billion Yuan (over US$ 480 million) to help state-owned and small local coal mines in gas explosion prevention and monitoring.

In 2003 in all coal mines 4,143 fatal accidents occurred and 6,434 miners died. State-owned coal mines realized the drop of fatal accidents and fatalities. Altogether in state-owned coal mines 1,138 fatal accidents occurred and 1,773 miners died, comparing to 2002 the number of accidents decreased by 37 and the number of fatalities by 154, 3.15% and 7.99% respectively. Among the fatal accidents and fatalities, 531 fatal accidents occurred in state-owned key coal mines and 892 miners died, comparing to 2002 the number of accidents increased by 15 and the number of fatalities dropped by 12, respectively increased by 2.90% and dropped by 1.33%; 607 fatal accidents occurred in local state-owned coal mines and 881 miners died, comparing to 2002 the number of accidents dropped by 52 and the number of fatalities by 142, 7.89% and 13.88% respectively. The accidents and fatalities which occurred in state-owned coal mines in 2003 make up 27.47% and 27.55% respectively of those in all coal mines. The numbers of fatalities in state-owned coal mine accidents from 2001 to 2003 are shown in figure 3 and the fatality rate per one million ton of output in state-owned coal mines in figure 4 (SACMSS Web 2004, SAWS Web 2004).

It's predicted that China's coal output in 2004 will reach above 1.9 billion tons, twice of that in 2000, but the fatality rate of every one million tons of output will be brought to under three, lower than 5.77 in 2000 (Zhao et al. 2004).

Economic losses caused by fatal accidents and occupational diseases are extremely huge. For every accident coal mines must pay considerable amount of money for rescue, medical treatment, comforts and aids to the affected families, children and so on. Especially gas explosion accidents additionally destroy engineering facilities and equipments, resulting in bigger direct and indirect losses. According to data of some mining bureaus, every accident with 1-fatality results in direct and indirect loses of about 300,000 Yuan (Peng at el. 2002); the annual economic loss reaches nearly 10,000 Yuan for a pneumoconiosis patient (including treatment and working ability loss). According to the standards, for all coal mines because of accidents and occupational diseases annual economic loss reaches as high as nearly 4 billion Yuan, and is equal to about 10% of annual coal sale income of state-owned key coal mines (Peng at el. 2002).

Therefore, improvement of work safety and work conditions in coal mines always is an important issue in the Chinese coal industry.

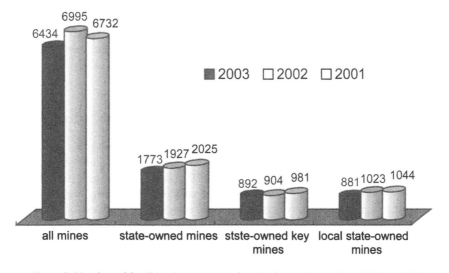

Figure 3. Number of fatalities in state-owned coal mine accidents from 2001 to 2003

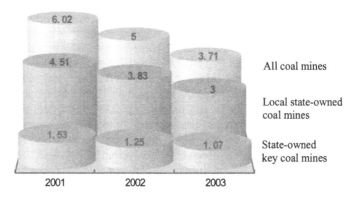

All coal mines

Local state-owned
coal mines

State-owned
key coal mines

Figure 4. Fatality rate per one million ton output in state-owned coal mines from 2001 to 2003

4. INFLUENCE OF COAL MINING ON ENVIRONMENT

4.1. *Land surface destruction caused by coal mining*

Underground mining results in land surface destruction. At present in China 95% coal output come from underground mining, moreover in state-owned key coal mines basic mining method is longwall mining with roof caving system. In this method the greatest depth of land surface subsidence is generally 0.7 times the total thickness of the mined out coal bed, the influence area is about 1.2 times the total thickness of the mined out coal bed. According to incomplete statistics, land surface destruction area caused by coal mining in China reaches 400,000 hectares, meaning that every ten thousand tons of coal output results in 0.2 hectare land surface destruction. Obviously land surface destruction caused by coal mining is very serious (Huang 2003).

4.2. *Influence of coal mining on water resources*

One of the influences of coal mining on the environment is destruction of water resources, in particular underground water resources. Although at present there are no statistics of damage to national water resources by coal mining, a report indicated that in Shanxi Province water resources destruction by coal mining is quite serious, equally every ton of coal output causes 2.5 tons of water loss (Resources & Environment Web 2004).

4.3. *Influence of mine gas emission on atmosphere*

Air pollution caused by coal mining process is mainly due to emission of mine gas (principal constituent is methane) and gas released in gangue spontaneous combustion. Due to the fact that the greenhouse effect caused by it is 21 times that of CO_2, methane is also considered one of the main gases causing global climate changes. According to statistics from the State Administration of Coal Mine Safety Supervision, in China in 2001 methane drainage reached 980 million cubic meters (pure CH_4 equivalent), of which less than 50% was used. The quantity of gas released to the atmosphere from drainage exceeded 500 million cubic meters. This not only wastes a lot of clean energy, but also causes serious pollution to atmospheric environment (Huang 2003).

4.4. *Influence of gangue spontaneous combustion on atmosphere*

Gangue spontaneous combustion produces poisonous gas containing SO_2, CO_2, CO, which is another major bad influence of coal mining on the atmospheric environment. According to statistics from the State Administration of Coal Mine Safety Supervision, in China at present state-owned coal mines altogether have 1,500 gangue stacks, among which 389 gangue stacks in the state of spontaneous combustion for a long time, which have seriously polluted atmospheric environment in the mining area and its peripheral locality (SACMSS Web 2004).

4.5. *Influence of coal direct combustion on environment*

In China coal consumption is mainly by raw coal direct combustion. Only approximately 38% of raw coal is processed, comparing with that of main coal producers in the world the difference is 20~60%, but 62% of raw coal without washing and processing is directly used for burning. In national total coal consumption, 85% of coal is burned for power generation, in industrial boilers, and for cooking. Raw coal direct combustion with low combustion efficiency has caused serious destruction of environment and results in occurrence of acid rain. Acid rain affects above 1/3 territory of China, the economic loss has reached as high as 2% GDP.

5. SOME TACTIC POLICY PROPOSALS FOR COAL INDUSTRY SUSTAINABLE DEVELOPMENT

Development is the core of sustainable development. Economical development is one of the basic conditions for coordinative development of population, resources, environment and economy. Sustainable development requests to consider the needs of both current and future development. It shouldn't be a cost of development to sacrifice the descendants' benefits just to satisfy the contemporary needs. Therefore energy production must satisfy middle-term and long-term economical development. In China population is large and average energy source possession per capita much lower than the average level in the world. Although coal resources can satisfy the needs for short-term economical fast development, energy supply must guarantee the long-term economical development, which can't depend on coal only. It's necessary to save coal resources and develop other energy sources, especially renewable energy. It is also important to obtain energy sources from the world energy market and establish a national energy reserve system to solve energy security problem. One of the important aspects of sustainable development is sustainable use of resources and keeping good ecological standards. Sustainable development requests to consider pollution control in energy development, transformation and use processes. It's necessary to decrease proportion of coal in the energy structure and improve combustion efficiency to satisfy environmental protection demands. In summary, the following policy proposals can be beneficial to the Chinese coal industry sustainable development:
1. Limit small coal mines and strengthen coal resources protection.
2. Improve coal mining technical equipments and technology to enhance coal recovery rate.
3. Improve coal mining methods to strengthen land and water resources protection.
4. Take measures to raise safety level in coal mines, reduce fatal accident occurrences.
5. Save energy, enhance coal use efficiency.
6. Develop renewable energy to optimise energy structure and reduce coal proportion in the energy structure.
7. Encourage using and developing clean coal, namely clean coal products, produce clean coal and cleanly use coal.
8. Control environmental pollution caused by coal combustion, especially control emission of pollution from coal-fired power plants.
9. Establish mature energy consumption market environment and coal marketing mechanism.
10. Take measures to solve energy security issue.

REFERENCES

China Web 2004: http://www.china.org.cn/english/en-shuzi2004/zr/zrzy-kc.htm

Fushun Branch of China Coal Research Academy 2003: 50 Years' Work at Fire Prevention and Treatment. Safety in Coal Mines. Vol. 34 additive issue, Sept. 2003.

Henan Coal Web 2004: http://www.hnmt.gov.cn

Huang S. 2003: Research on Influences of Coal Mining And Usage on Environment in China. Beijing, Aug. 2003.

Luo H., Liang Y. 2003: Current Status and Perspective of Forecast and Prediction Techniques of Spontaneous Combustion of Coal. China Safety Science Journal. Vol. 13, No. 3, March 2003.

Peng Cheng, Li Fenxi 2002: Accident Costs, Enterprises Should be Startled.
http://www.zgmt.com.cn/2002/ddkg2002/ddkg7/AQGL7sgdj.htm

Resources & Environment Web 2004: For One Ton Output of Coal Lose 2.5 Tons Water:
http://www.myearth.com.cn/MESSAGE/20020604/content_xy.asp

State Administration of Coal Mine Safety Supervision (SACMSS) Web 2004:
http://www.chinacoal-safety.gov.cn

State Administration of Coal Mine Safety Supervision (SACMSS) 2002: China Coal Industry Yearbook 2002. Beijing, May 2002.

State Administration of Work Safety (SAWS) Web 2004: http://www.chinasafety.gov.cn

Safety Committee of China Coal Society 1995: New Progress in Coal Mine Safety Technology in Our Country. Safety in Coal Mines. No. 1, 1995.

Wang X., Zhang G. 1994: The Current Situation and Achievements of Mine Fire Prevention and Treatment Techniques in China. Fire Safety Science. Vol. 3, No. 2, Sept. 1994.

Zhao X., Jiang X. 2004: Coal Mining: Most Deadly Job in China. China Daily Web, 2004.11.13, 15:01 at:
http://www.chinadaily.com.cn/english/doc/2004-11/13/content_391242.htm

International Mining Forum 2005, Sobczyk & Kicki (eds) © 2005 Taylor & Francis Group, London, ISBN 0415 375525

A Life Cycle Assessment of Electricity Production from Hard Coal

Jyri Seppälä
Finnish Environment Institute. Helsinki, Finland

Laura Sokka
Finnish Environment Institute. Helsinki, Finland

Sirkka Koskela
Finnish Environment Institute. Helsinki, Finland

Joanna Kulczycka
Polish Academy of Sciences, Mineral and Energy Economy Research Institute. Cracow, Poland

Malgorzata Goralczyk
Polish Academy of Sciences, Mineral and Energy Economy Research Institute. Cracow, Poland

Karol Koneczny
Polish Academy of Sciences, Mineral and Energy Economy Research Institute. Cracow, Poland

Anna Henclik
Polish Academy of Sciences, Mineral and Energy Economy Research Institute. Cracow, Poland

ABSTRACT: Since coal is a fossil fuel used commonly for electricity production in the world, life cycle assessments (LCAs) of products and services usually require data on the environmental burdens of electricity production from coal. In an EU Life-Environment project called OSELCA, one task was to create LCA data on electricity production from hard coal and oil shale in order to compare the environmental performances of these two electricity sources. For this purpose, an LCA of hard coal electricity was carried out. The aim of this article is to give a general view of the emissions and environmental impacts caused by the various life cycle stages of hard coal electricity on the basis of the LCA study. The production system consists of hard coal mining in Poland, a power plant representing the best available technology in Finland, and all the main processes related to mining and the power plant. The emissions of the electricity system were assessed and analysed from the point of view of different impact categories. For acidification, tropospheric ozone formation and terrestrial eutrophication, the country-dependent characterisation methodology of LCA was used. The results showed that the power plant stage is a major life cycle stage causing climate change, acidification, eutrophication and heavy metals to air (except for Hg). The coal mining causes 95% of the non-recovered wastes of the whole production system and it contributes most to the effects of tropospheric ozone formation due to methane emissions. Also external electricity and heat production for mining and coal transportation are important contributors to tropospheric ozone formation, acidification, eutrophication, particulate matter and some heavy metals (Cu, Ni and Zn) to water.

KEYWORDS: Coal, electricity, environmental impact assessment, life cycle assessment (LCA)

1. INTRODUCTION

Life cycle assessment (LCA) appears to be a valuable tool for assessing the environmental aspects of a product or a service throughout its entire life cycle. For this reason, an EU Life-Environment project[1] called *Introduction and Implementation of Life Cycle Assessment Methodology in Estonia: Effects of Oil Shale Electricity on the Environmental Performance of Products (OSELCA)* was launched in September 2003 in order to disseminate and promote life cycle thinking and the Integrated Product Policy (IPP) approach in Estonia. The OSELCA project consists of eight tasks altogether. The aim of Task 2 is to identify the differences between the environmental impacts caused by oil shale electricity and coal electricity through their whole life cycles. For this reason, an LCA of electricity production from hard coal was carried out in the project.

Estonia is the only country in Europe with a considerable oil shale mining industry. Over 95% of Estonian electricity generation is produced with oil shale. In Task 2 of the OSELCA project, the environmental impacts of oil shale electricity are compared with those caused by hard coal in order to get a better view about the environmental impacts of oil shale electricity generation. Hard coal was chosen for the comparison because it is a commonly used fossil energy source.

For the LCA of hard coal electricity, the Finnish Meri-Pori power plant representing the best available power plant technology was chosen as a case study power plant. Although Meri-Pori obtains its fuel from many different countries, coal-mining data were collected using the average data on hard coal mining in Poland because Poland is one of the largest suppliers of hard coal to Finland. Thus, this particular inventory does not represent any real production system as such but could, however, be theoretically possible.

The purpose of this article is to give an overview of the LCA of hard coal electricity. In this article, a brief summary of the inventory is presented. The preliminary results of LCIA are then addressed on the basis of the inventory.

2. MATERIAL AND METHODS

2.1. *Goal and scope of the study*

The main objective of this LCA study is to give a general view of the emissions and environmental impacts caused by the various life cycle stages of electricity production from hard coal. Moreover, the purpose of the study is to provide the methodological framework for the comparison between oil shale and coal electricity.

In this LCA, 1 MWh of electricity delivered to the consumer was used as a functional unit. The functional unit offers a reference unit for which the inventory and impact assessment results will be presented, making it possible to compare the results of oil shale electricity with the results of coal electricity. Finnish average transmission losses (3.5%), (Statistics Finland 2003) were included in the product system. The end use of electricity was, however, excluded from this inventory.

The study covers all the main processes in the life cycle of the product, starting from the mining of hard coal and ending at the delivery of electricity to consumers (figure 1).

---

[1] The leader of the OSELCA project is the Estonian state-owned energy company Eesti Energia AS and the partners are the Estonian consultancy CyclePlan Ltd. and the Finnish Environment Institute (SYKE). The Polish Mineral and Energy Economy Research Institute has participated in Task 2 of the project as an expert on coal mining issues.

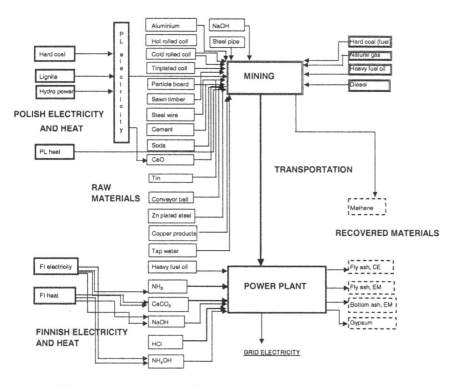

Figure 1. A product system of hard coal electricity and its life cycle stages

2.2. *Life cycle inventory analysis*

The LCI of hard coal electricity was conducted according to the recommendations of the International Organisation for Standardisation (ISO 14040, 1997, ISO 1401, 1998). The product system was described as a flowchart and the LCI results were calculated with the help of the KCL-ECO 3.02 software (Keskuslaboratorio Oy 2003). For the unit processes, which are the smallest units for which information is gathered in an LCI, data on their inputs (energy, materials, land use) and outputs (emissions, wastes, products, by-products) were collected. On the basis of the flowchart, modules containing one or several unit processes were divided into different life cycle stages. The division of the production system into different life cycle stages helps the interpretation of the LCI results. For this inventory the following main life cycle stages are used (see figure 1):

1) Raw materials for a) coal mining and b) the power plant.
2) External electricity and heat production in Poland (coal mining and its raw materials).
3) External electricity and heat production in Finland (power plant and its raw materials).
4) Hard coal mining and processing.
5) Transportation:
   a) electric train transportation of coal in Poland (including interventions caused by Polish electricity production);
   b) sea transportation of coal from Poland to Finland, and
   c) other transportation.
6) Hard coal power plant.
7) Recovered wastes treated as by-products:
   a) in Poland, and
   b) in Finland.

Under the life cycle stages 1 and 7, two sub-stages were considered. For transportation, three sub-groups were applied. The sub-division offered a better basis to arrange the data according to the different geographical areas utilized in the impact assessment (see Section 2.3).

For hard coal mining, average Polish data were used and they were provided by Polish Mineral and Energy Economy Research Institute. Also for the Meri-Pori power plant, site-specific data were received from the company itself. For the production of the raw materials and fuels, secondary data from different databases, mainly Ecoinvent (Swiss Centre for Life Cycle Inventories 2004), KCL-Ecodata (Keskuslaboratorio Oy 2003), IISI LCI database (International Iron and Steel Institute 2002) and LIPASTO (Technical Research Centre of Finland, VTT 2003), were used. The environmental interventions of external electricity production in Poland and in Finland were calculated on the basis of the shares of different fuels in the electricity production profiles of these two countries. In the case of Polish electricity, the environmental interventions from producing these fuels were added to the Polish electricity model from the Ecoinvent database representing Polish conditions. In the case of Finland, the results of a Finland-specific electricity model developed by the Finnish Environment Institute (Laukka 2003) were used.

Transportation of raw materials was included in the inventory whenever it was possible. Indirect (from production of the fuels) and direct emissions together constituted the output factors used in the transportation modules. Coal was assumed to be transported from Gdansk harbour to Meri-Pori (810 km in one direction).

Some by-products were created during the mining and the power plant stages, namely methane from mining, and fly ash, bottom ash and gypsum from the power plant. Most of the mining waste was re-used for industrial purposes, but as it was not possible to obtain accurate information over its utilisation. The 'avoided burden approach' was adopted for the treatment of the by-products. This means that credits were given for recycled wastes. This was done by deducting the inputs and outputs of producing the quantity of the material that the by-product replaces from the total results for hard coal electricity. The idea is that the by-product replaces some other product in another product system, and hence emissions from the saved processes are avoided. Recycling of steel was treated according to a methodology developed by the International Iron and Steel Institute (2002). In the case of multi-product systems, inputs and outputs were allocated to 'expanded system boundaries' (ISO 1998). Since this inventory dealt with only one power plant, which produces solely electricity, no allocation – i.e., division of inputs and outputs between different forms of energy – was needed.

In the inventory, the data on mining were for the year 2002 and data on the power plant for the year 2003. The latest possible year was also used for all other modules and unit processes.

2.3. *Life cycle impact assessment*

In this study, the inventory data were analysed using the classification and characterisation methodology of LCIA (ISO 14042, 2000). First, appropriate impact categories (e.g., climate change and acidification) were selected on the basis of the existing inventory data and the general knowledge about cause-effect relationships. After that the inventory data were assigned into the impact categories (classification). In the characterisation, the chosen characterisation factors enable an aggregation of the emissions within each impact category. The emission values are converted into impact category indicator results by multiplying the emission values by the corresponding characterisation factors. In order to produce scientifically based characterisation results, the determination of characterisation factors within a certain impact category is a key issue.

The following impact categories and environmental interventions causing the effects of the impact categories were identified:

1. climate change (CO_2, N_2O, CH_4);
2. acidification (SO_2, NO_x (expressed as NO_2), NH_3);
3. tropospheric ozone formation (NO_x, NMVOC);

4. aquatic eutrophication (P (to water), N (to water), NO_x, NH_3);
5. terrestrial eutrophication (NO_x, NH_3);
6. human toxicity (e.g. benzene, formaldehyde, toluene, metals to air);
7. ecotoxicity (e.g. dioxins, PAH, PCB, metals, oil, cyanides and phenols to water);
8. particulate matter ($PM_{2.5}$);
9. depletion of natural resources (oil, gas, coal, selected minerals);
10. effects of land use, and
11. wastes.

In the case of climate change, the global warming potential (GWP) factors of national greenhouse gas inventories were used as characterisation factors. For tropospheric ozone formation, acidification and terrestrial eutrophication, the latest country-specific characterisation factors were used instead of site-generic characterisation factors (tropospheric ozone formation: (Hauschild et. al. 2004), acidification and terrestrial eutrophication: (Seppälä et al. 2005)). This is due to the fact that the location of the emission source can cause different responses in the surrounding ecosystems in the context of these impact categories, depending, e.g., on local atmospheric conditions and the sensitivity of the ecosystems (see, e.g., (Amann et al. 1999)). In the context of these four impact categories, the emissions of life cycle stages 1a, 2, 4, 5a and 7a were calculated using characterisation factors specific to Poland, whereas life cycle stages 1b, 3, and 6 were calculated using characterisation factors specific to Finland. For sea transportation of hard coal from Poland to Finland (5b), characterisation factors specific to the Baltic Sea were used. Other transportations (5c) were calculated using the average factors for Poland and Finland.

Under tropospheric ozone formation, effects on human health and vegetation were calculated as separate impact categories on the basis of the RAINS model (Amann et al. 1999). In the method used, the assessment of health effects is based on weighted AOT60 (= the accumulated amount of ozone over the threshold value of 60 ppb) values as impact category indicators, whereas the effects on vegetation are based on the use of weighted AOT40 values. AOT60 values are weighted by population exposure and AOT40 values are weighted by the area of vegetation ecosystems in Europe. In acidification and terrestrial eutrophication, critical loads are taken into account, using accumulated exceedance as category indicators. The country-dependent characterisation factors of both impact categories are based on the use of an integrated assessment model for Europe developed by EMEP (1998).

The impacts on aquatic eutrophication were calculated according to a rough characterisation method because it is well known that power plants do not cause a significant direct nutrient flux to water. For this reason, the nutrient emissions were calculated using the same characterisation factors for all the life cycle stages. The weakness of this method is that the characterisation factor for nitrogen oxides to air was only adapted for Finnish emissions. It was also assumed that half of the nitrogen and phosphorous contents of the total nitrogen and phosphorous emissions to water, and all waterborne inorganic nitrogen and phosphate phosphorous emissions, cause aquatic eutrophication (see (Seppälä et al. 2004)).

For the impact categories of human toxicity, ecotoxicity and particulate matter, characterisation was not conducted. This is due to the fact that the complete inventory data and/or the reliable characterisation factors for those impact categories, taking into account the conditions of Northern and Central Europe, are practically not available. For this reason, the contributions of various life cycle stages to those impact categories were only analysed by arranging the emissions according to the life cycle stages. Waste generation was handled in the same way.

The effects of land use and depletion of natural recourses are omitted in this article. In this stage of the project, data along the life cycle stages are not reliable enough to make these assessments.

3. RESULTS AND DISCUSSION

In the following, the results are presented according to impact categories. A detailed list of all the environmental interventions of the product system used for analysing the impact categories can be found in the Appendix.

Most of the emissions causing *climate change* occur at the power plant (figure 2). It causes 95% of the CO_2 emissions in the production system. The methane emissions from mining cause about 4% of the global warming potential of the whole production system. The contribution of the other life cycle stages is insignificant.

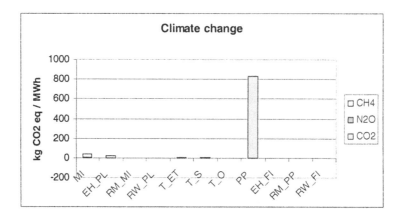

Figure 2. Contributions of different emissions and life cycle stages to climate change.
Abbreviations used for the life cycle stages: MI – hard coal mining and processing;
EH_PL – external electricity and heat production in Poland (coal mining and its raw materials);
RM_MI – raw materials to coal mining; RW_PL – recovered wastes treated as by-products in Poland;
T_ET – electric train transportation in Poland (including emissions caused by electricity production in Poland);
T_S – sea transportation of coal from Poland to Finland; T_O – other transportation; PP – power plant;
EH_FI – external electricity and heat production in Finland (for the power plant and its raw materials);
RM_PP – raw materials to the power plant; RW_FI – recovered wastes treated as by-products in Finland

The power plant causes most of the NO_x and SO_2 emissions (Appendix). However, the difference between country-specific characterisation factors for *acidification* increased the relevance of transportation and electricity production in Poland in the final impact category indicator results (figure 3). In practice, SO_2 contributes most to acidification.

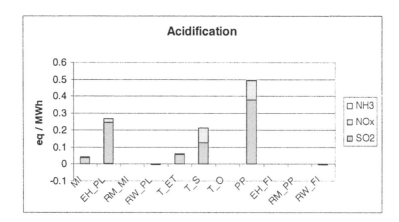

Figure 3. Contributions of different emissions and the life cycle stages to acidification

18

In *tropospheric ozone formation*, the methane emissions of mining cause the greatest effects on human health (figure 4). The NO_x emissions of the power plant situated in Finland do not cause health effects because of low population density and background ozone concentration, and the unsuitable environmental conditions for ozone formation in Northern Europe. In the case of effects on vegetation, the NO_x emissions of the power plant in Finland make a larger contribution, although methane from mining still plays the greatest role in these effects. In the interpretation it should be kept in mind that the figures of methane include large uncertainty (see Hauschild et al. 2004).

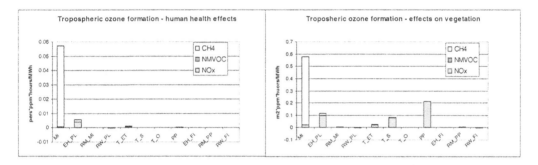

Figure 4. Contributions of different emissions and the life cycle stages to effects on human health and vegetation caused by tropospheric ozone formation

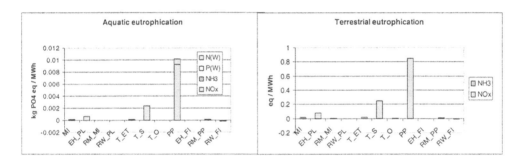

Figure 5. Contributions of different emissions and life cycle stages to aquatic and terrestrial eutrophication

In aquatic *eutrophication*, the contribution of NO_x emissions dominates compared with waterborne nutrient emissions (figure 5). In practice, the contributions of the various life cycle stages correspond directly to the NO_x emissions of those stages. Due to the rough characterisation method, this conclusion is subject to uncertainty. The contribution of NO_x emissions to *terrestrial eutrophication* is similar to its contribution to aquatic eutrophication.

Fine particles are the most harmful for human health. In the inventory, primary particles smaller than 2.5 µm, $PM_{2.5}$, were assessed. The power plant and external electricity and heat production for mining are the biggest contributors to $PM_{2.5}$ emissions (figure 6). In terms of these pollutants, also the production of limestone for the power plant plays an important role. However, in order to create a complete overview about the contribution of the different life cycle stages, inorganic emissions of secondary particles (SO_4, NO_3, NH_4) and secondary fine particles (sPM) should be assessed. In addition, characterisation factors taking into account spatial differences in population density should be used.

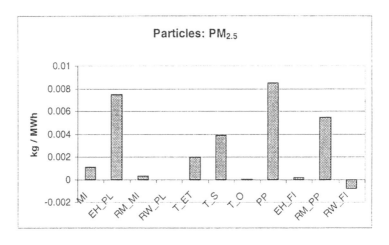

Figure 6. PM$_{2.5}$ emissions (kg/MWh) according to the life cycle stages

Under *human toxicity and ecotoxicity*, only the results for heavy metals are presented. This is because that in this stage of the project it has not yet been possible reliability to assess organic harmful compounds for all the life cycle stages. Most of the heavy metal emissions to air occur at the power plant (figure 7). The power plant causes 95−57% of all the other metal emissions except for Hg, of which it causes 11%. Polish electricity production (mainly used in the mining operations) is another relevant contributor to the heavy metal emissions. It causes most (~63%) of the Hg emissions and 5−25% of the other metal emissions. Transportation of hard coal within Poland by electric trains also generates about 17% of the Hg emissions. In the case of waterborne heavy metals, the external electricity and heat production for mining processes and electric train transportation causes clearly the largest heavy metal releases of Cu, Ni, and Zn, and their absolute values expressed in grams are in a different order of magnitude compared to the values of all other metals except for Fe (see Appendix).

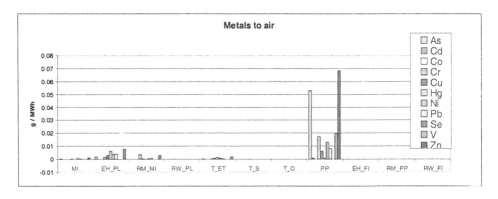

Figure 7. Total emissions to air of heavy metals (g/MWh)

The largest waste fraction of the system, 109.1 kg/MWh, is created during the processing of hard coal (table 1). Mining (excavation) generates about 7.5 kg waste/MWh. Almost 90% of the mining waste is recovered for engineering works. Also the flying ash and gypsum produced at the power plant are large waste fractions. Over 99% of the waste produced by the power plant is recycled. The waste generation during the raw material production is minimal compared to the waste from mining and the power plant. Additionally, the amount of avoided waste from the saved external processes is larger than the amount of waste generated by raw material production.

Table 1. Waste generation according to life cycle stages. The table includes total amounts without recovery. Almost 90% of the mining waste and over 99% of the waste caused by the power plant is recovered

WASTE FRACTION	AMOUNT (kg/MWh)
MINING	
Waste from processing	109.1
Waste from excavation	7.5
Flotation waste	7.3
Other waste	0.3
POWER PLANT	
Fly ash	34.0
Bottom ash	5.1
Gypsum	8.9
Other waste	0.1
RAW MATERIAL PRODUCTION	
Total waste	0.04
SAVED EXTERNAL PROCESSES	
Mineral waste, inert	-0.05
OTHER	
Total hazardous waste from all life cycle phases	0.01

4. CONCLUSIONS AND FUTURE OUTLOOK

According to this LCA of hard coal electricity, the power plant causes over half of the total impacts of the whole electricity production system in the case of climate change (92%), aquatic eutrophication (71%) and terrestrial eutrophication (61%). The power plant seems also to be the most important source of heavy metals to air, causing 95%−57% of all other metal emissions except mercury. External electricity and heat production used by the production system in Poland cause a major part of the airborne Hg. It also causes clearly the largest waterborne heavy metal releases of Cu, Ni and Zn. In acidification, the power plant is the biggest contributor, but external electricity and heat production for mining and the sea transportation of coal also have large roles in causing acidification. In the case of tropospheric ozone formation, methane from mining contributes over 80% of the total human health effects caused by the whole production system. In addition, the contribution of the same life cycle stage to effects on vegetation is over half of the total ozone impacts caused by the production system. In the case of primary fine particles, the power plant external causes 30% of $PM_{2.5}$ emissions of the whole life cycle. Furthermore, the roles of external electricity and heat produced for mining (27%) and the production of limestone to the power plant (20%) are significant for $PM_{2.5}$ emissions over the whole life cycle.

Credits due to the reuse of waste materials have minor relevance in the results of the whole life cycle except for particles, which were reduced by almost 3% through saved limestone production. Processing of hard coal in mining causes the largest amounts of waste in the life cycle (app. 70% of the total non-recycled waste amount). In practice, over 99% of the waste fractions of the power plant are utilised.

In the use of the LCA results, it should be kept in mind that the results for electricity production from hard coal represent the data sources related to certain technology levels/choices. For example, the final emissions of NO_x, SO_2, particulate matters and heavy metals depend on the quality of the hard coal used and on the air emission purification systems used. The coal-fired power plant of this LCA represents a very high level of technology (BAT) including desulphurisation and selective catalytic reduction of NO_x emissions from the flue gas (see (Finnish Environment Institute 2001)). The data on the electricity and heat production in Poland were derived from the Ecoinvent databases, and it is unclear how well the data represent the current average power plant technologies in Poland.

The data of the LCA study provide a basis for further applications on hard coal electricity in which there is a need to choose other assumptions (e.g., distances for coal transportation) and unit processes/modules to better describe the desired production system. In addition, the country-dependent characterisation factors used in the study offer a new approach to the interpretation of environmental impacts.

In the future[2], there is a need to conduct sensitivity analysis to check how the results of this LCA change when the inventory data are varied by factors including uncertainty or representing a different technological choice (e.g., using the average emissions of power plants in Finland instead of using the emissions of BAT). There is also a plan to complement the inventory with data on land use, natural resources and harmful substances causing toxicity in order to conduct the impact assessment for human toxicity, ecotoxicity, depletion of natural resources and the effects of land use. In addition, the normalisation stage of life cycle impact assessment (see ISO 14042, 2000) will be carried out in order to provide a more comprehensive picture of the relative importance of the different impact categories.

REFERENCES

Amann M., Cofala J., Heyes C., Klimont Z. and Schöpp W. 1999: The RAINS Model: A Tool for Assessing Regional Emission Control Strategies in Europe. Pollution Atmosphérique 20, 41−46.

EMEP 1998: Transboundary Acidifying Air Pollution in Europe. EMEP/MSC-W Report 1/98. Norwegian Meteorological Institute, Oslo.

Finnish Environment Institute 2001. Finnish Expert Reports on Best Available Techniques in Large Combustion Plants. The Finnish Environment 458. Finnish Environment Institute, Helsinki.

Hauschild M., Bastrup-Birk A., Hertel O., Schöpp W. and Potting J. 2004: Photochemical Ozone Formation. In: Potting J. and Hauschild M. (eds.): Background for Spatial Differentiation in Life Cycle Assessment − the EDIP 2003 Methodology. Institute of Product Development, Copenhagen.

International Iron and Steel Institute, IISI 1998. LCI Database.

International Iron and Steel Institute, IISI 2002. World Steel Life Cycle Inventory. Methodology Report 1999/2000. Committee on Environmental Affairs. IISI, Brussels.

ISO (International Organization for Standardization) 1997: ISO 14040: Environmental Management − Life Cycle Assessment − Principles and Framework. International Organization for Standardization, Geneva.

ISO (International Organization for Standardization) 1998: ISO 14041: Environmental Management − Life Cycle Assessment − Goal and Scope Definition and Inventory Analysis. International Organization for Standardization, Geneva.

ISO (International Organization for Standardization) 2000: ISO 14042: Environmental Management − Life Cycle Assessment − Life Cycle Impact Assessment. International Organization for Standardization, Geneva.

Keskuslaboratorio Oy 2003: The Finnish Pulp and Paper Research Institute. KCL-ECO 3.02. Available on-line at: http://www.kcl.fi/eco/indextxt.html

Laukka J. 2003: Life Cycle Inventories of Newspaper with Different Waste Management Options. A Case Study in the Helsinki Metropolitan Area. M.Sc. Thesis. Department of Environmental Sciences, University of Kuopio.

---

[2] The OSELCA project will last until the end of 2005. For more detailed descriptions of the methods, data sources and results see the full report on Task 2, which will be available on the OSELCA website (www.energia.ee/OSELCA) in April, 2005.

Seppälä J., Knuuttila S. and Silvo K. 2004: Eutrophication of Aquatic Ecosystems. A New Method for Calculating the Potential Contributions of Nitrogen and Phosphorus. Int J LCA 9 (2) 93–100.

Seppälä J., Posch M., Johansson M. and Hettelingh J.P. 2005: Country Dependent Characterization Factors for Acidification and Terrestrial Eutrophication Based on Accumulated Exceedance as Impact Category Indicator. Submitted to Int J LCA.

Statistics Finland 2003: Energy in Finland 2002. Helsinki: Statistics Finland.

Swiss Center for Life Cycle Inventories 2004: Ecoinvent Database, v. 1.01 & 1.1.

Technical Research Center of Finland, VTT 2003. LIPASTO Database. Lipasto Database of Traffic Emissions. VTT Technical Research Centre of Finland.

APPENDIX

Emissions (kg) per MWh according to different life cycle stages

	MI	EH_PL	RM_MI	RW_PL	T_ET	T_S	T_O	PP	EH_FI	RM_PP	RW_FI
CO_2	3.16	19.98	0.41	-0.07	4.44	6.53	0.06	831.78	0.34	0.57	-1.65
N_2O	2.30E-05	1.35E-04	1.15E-05	-6.56E-07	2.67E-05		2.47E-06		2.71E-06	4.86E-07	-2.87E-05
CH_4	1.68E+00	5.25E-02	7.07E-04	-2.00E-03	1.16E-02		7.10E-06		8.62E-04	3.21E-03	-1.75E-03
SO_2	2.10E-02	1.42E-01	1.18E-03	-9.83E-04	3.24E-02	1.05E-01	1.10E-05	8.22E-01	6.25E-04	7.37E-04	-2.34E-03
NO_x	8.78E-03	3.91E-02	1.35E-03	-1.91E-04	8.39E-03	1.55E-01	6.50E-04	6.02E-01	7.07E-04	8.89E-03	-5.68E-03
NH_3			5.40E-08							2.09E-04	-1.43E-06
NMVOC	3.96E-04	5.69E-04	1.84E-04	-6.63E-04	1.03E-04	5.14E-03	8.84E-05		5.85E-04	1.70E-04	-1.16E-04
$PM_{2.5}$	1.15E-03	7.47E-03	2.74E-04	-1.25E-05	1,73E-3	3.88E-03	6.33E-06	8.52E-03	1.69E-04	5.49E-03	-7.86E-04
As	3.02E-07	2.11E-06	5.79E-08	-7.01E-10	3.84E-07			5.29E-05	5.93E-10	2.91E-09	-6.31E-10
Cd	3.07E-08	2.36E-07	1.50E-08	-3.20E-10	4.92E-08			8.56E-07	4.17E-10	1.09E-09	-4.25E-10
Co			3.18E-13								
Cr	1.66E-07	1.69E-06	3.48E-06	-4.84E-09	2.89E-07			1.74E-05	2.72E-09	4.76E-08	8.84E-10
Cu	3.36E-07	3.20E-06	3.64E-07	-5.55E-09	6.95E-07			6.34E-06	1.70E-09	1.35E-08	-9.48E-09
Hg	1.31E-07	6.17E-06	1.11E-07	-4.27E-09	1.51E-06			9.82E-07	2.07E-09	6.38E-09	-4.75E-10
Ni	6.23E-07	4.05E-06	2.85E-07	-5.51E-09	8.41E-07			1.33E-05	8.74E-09	1.73E-08	-6.05E-09
Pb	5.12E-07	3.96E-06	8.84E-07	-1.15E-08	5.65E-07			8.02E-06	8.33E-09	1.27E-08	5.39E-09
Se			5.54E-11								
V			6.28E-12					1.96E-05	2.39E-08		
Zn	1.23E-06	7.82E-06	3.11E-06	-3.05E-08	1.90E-06			6.85E-05	3.73E-09	2.67E-08	-1.32E-08
As (w)[1]	1.04E-05	5.50E-05	1.31E-06	-2.68E-08	1.29E-05			1.20E-07		1.44E-08	-6.68E-09
Cd (w)	7.87E-08	3.69E-06	1.01E-06	-1.82E-08	9.08E-07			1.90E-08		3.68E-09	-3.21E-09
Cr (w)	4.42E-08	1.61E-07	1.21E-07	-3.02E-09	3.80E-08			1.00E-06		1.17E-09	-1.34E-09
Cu (w)	2.48E-06	1.39E-04	8.17E-05	-1.23E-07	3.43E-05			8.71E-07		2.14E-07	-6.93E-08
Fe (w)	1.29E-04	2.86E-02	1.51E-03	-3.95E-06	7.08E-03			9.51E-06	1.12E-09	1.48E-05	-2.07E-04
Hg (w)	1.00E-08	8.01E-07	1.02E-07	-2.02E-09	1.98E-07			1.49E-07		1.22E-09	-3.32E-10
Mg (w)			2.96E-15								
Mn (w)			1.63E-14								
Ni (w)	9.44E-06	1.53E-04	4.24E-05	-3.29E-07	3.73E-05			1.65E-7			
Pb (w)	7.28E-07	1.46E-05	7.11E-06	-6.18E-08	3.62E-06			6.02E-7		2.91E-08	
V (w)								2.20E-06			
Zn (w)	3.35E-06	1.63E-02	2.10E-05	-5.02E-07	4.06E-03			1.77E-06		5.44E-06	-1.33E-06
Tot P(w)	2.19E-08	6.49E-08	4.54E-07	-2.99E-10	1.34E-08			2.06E-06	1.77E-08	6.36E-10	-2.38E-09
PO_4 (w)[2]			3.56E-7								-1.20E-5
Tot N (w)	8.77E-07	4.06E-05	2.09E-06	-3.38E-08	9.84E-06	3.75E-06	3.29E-08	1.37E-03	4.91E-07	1.60E-06	-8.10E-08
NH_4, NH_3 (w)[3]	2.15E-06	8.97E-06	4.55E-06	-4.45E-08	2.00E-06			1.32E-03	2.08E-07	1.34E-06	-3.26E-06
NO_2, NO_3 (w)[3]	1.20E-05	4.60E-05	9.64E-07	-3.39E-08					3.08E-08	1.09E-06	-9.18E-07

Notes: [1](w) — emission to water; [2]expressed as P, and [3]expressed as N.

Abbreviations used for the life cycle stages: MI — hard coal mining and processing; EH_PL — external electricity and heat production in Poland (coal mining and its raw materials); RM_MI — raw materials to coal mining; RW_PL — recovered wastes treated as by-products in Poland; T_ET — electric train transportation in Poland (including emissions caused by electricity production in Poland); T_S — sea transportation of coal from Poland to Finland; T_O — other transportation; PP — power plant; EH_FI — external electricity and heat production in Finland (for the power plant and its raw materials); RM_PP — raw materials to the power plant; RW_FI — recovered wastes treated as by-products in Finland.

International Mining Forum 2005, Sobczyk & Kicki (eds) © 2005 Taylor & Francis Group, London, ISBN 0415 375525

Scientific Problems of Coal Deposit Development in Ukraine

Gennadij G. Pivniak
National Mining University. Dnepropetrovsk, Ukraine

Volodymyr I. Bondarenko
National Mining University. Dnepropetrovsk, Ukraine

Petro I. Pilov
National Mining University. Dnepropetrovsk, Ukraine

Ukraine, for the different reasons of subjective and objective character has put itself in a position of countries with energy deficiency. Its own sources cover only 50% of its needs for fuel and energy resources. The majority of energy sources, first of all gas and oil, is compelled to come from import. Own extraction covers only 10–12% of demand for oil and 20–25% for natural gas. In fuel and energy balance of Ukraine the part of own energy carriers (including coal) has decreased to critically low level. In this difficult situation a new approach to strategy of the fuel and energy complex of the country is necessary.

Coal as energy source is unique for Ukraine. Its reserves are sufficient for some hundreds of years. Deposits with rather favorable conditions for use of modern technologies can provide the effective work of the industry for 70–80 years. It allows to provide for the economy's demand for coal due by extraction of the Ukrainian coal at competitive pries in the domestic market.

Balance coal deposits of A+B+C$_1$ categories in Ukraine amount to about 45 billion t., including above 13 billion t. of coke coal. About 1 billion t. is suitable for opencast mining. The basic coal deposits are concentrated in the Donetsk basin (up to 90%).

About 23 billion t of coal A+B+C$_1$ categories is ready prepared for industrial development. On balance of the working enterprises there are about 9,5 billion t. (including steam coal – 5,5 billion t., 58%).

However, coal deposits of Ukraine and, first of all, Donbass, are characterized by difficult geological conditions: low thickness of seams, great depth, weak stability of surrounding rock, high gas content, propensity to gas outbursts, mining impacts and other. About 77% of collieries are classified as gassy.

The coal seams with thickness greater than 1,2 m comprise only 20,4% of the total, coal seams with thickness up to 1,2 m – 74%, including coal seams with thickness smaller than 0,8 m) – – 33,3%.

Coal deposits of Ukraine contain significant quantities of methane. By various estimations the common resources of methane in Donbass make from 2,5 up to 25 bln. m^3. Thus, coal deposits of Ukraine should be considered as containing coal and gas.

In Donetsk basin there are 45 available sites, which can be recommended for constructing of 48 new mines. However 10 of them will have depths over 1400 m. Design studies have shown that practically all the mines planned for construction will face difficult geological conditions.

There are over 40 small fields of highly profitable mines superficial construction with total capacity 4–5 million t. per year.

Brown coal in Ukraine is also an essential fuel resource. Its use is limited due to the absence of boilers adapted to its use.

The total demand of Ukraine for commodity coal according to various scenarios of economy development amounts to: 2005 − 75,9–89,3 million t.; 2010 − 78–94 million. t.; 2015 − 80–100 million t.; 2020 − 82,2–110,5 million t.; 2025 − 90,1–120 million. t.; 2030 − 120–123,4 million t. Taking into account an exclusive role of coal in production of electric power, ferrous metallurgy, and in the developed energy strategy of Ukraine its development is very important.

The strategy of coal development provides increase of volumes and efficiency of own coal use as means of maintenance of power safety of Ukraine.

The major principles of strategy of the coal industry are:
- Structure of fuel and energy balance is directed on increase of volumes of coal consumption.
- Maintenance of internal demand for coal due to own coal industry.
- Improvement of quality and the level of state regulation in the branch by creation of effective control systems of state property.
- Stimulation of use of the domestic equipment and technologies of new generation.
- Creation of the state innovative and investment policy in coal industry.

Strategy considers, that the basic internal areas to increase efficiency of the coal-mining enterprises are:
- Level of capacities usage.
- New techniques and adequate technologies.
- Increase of an energy potential of the extracted coal due to optimum quality of the coal production received by processing of coal at processing plants with modern technologies.
- Use of secondary resources of fuel recovered from waste products.
- Change of activity of the coal-mining enterprises, transition to versatile manufacture.

Potential of the coal industry is very high: the prepared reserves of coal, an opportunity of coal mechanical engineering, new technologies of extraction, processing and use of coal, the advanced industrial infrastructure, highly skilled scientific, and the personnel which have an operational experience in difficult geological conditions.

Positions of strategy and the analysis of the coal industry of Ukraine follows, that new capacities will be created both due to construction of new mines, and due to reconstruction of working minefields by expanding and adapting existing infrastructures under new capacities.

Ordering of the technological problems arising during development of coal deposits of Ukraine, has allowed to formulate scientific and technical problems (fig. 1) solution of which will allow to reach an overall purpose of the strategy − increase the volumes of coal extraction maintaining the competitiveness of coal in the domestic market.

The geological conditions of Donbass, in particular such as low thickness of seams, big depth, high gas content, called for urgency in solving scientific and technical problems of the mining science, basic of which are:
- Studying of high mining pressure, gas outbursts phenomena, and temperature mode.
- Research of tensely deformable conditions of rockmass around the mining workings at big depths.
- Maintenance of underground mining workings by resource saving technologies on the basis of management of rockmass energy.
- Opportunities of creation of new techniques for development of low thickness seams.
- Creation of effective and safe technologies of coal output with use of new reliable and high-efficiency techniques.
- Creation of system of safety of mining operations adequate to modern conditions of coal deposits development.

These problems already constrain growth of volumes of extraction and labour productivity. For example, modern machinery (mechanized support sets, loaders, transport equipment, etc.) allowing to increase coal mining from one longwall up to 2000–6000 t/day are manufactured in Ukraine. However, their application conflicts with traditional systems of preparation and development of coal seams.

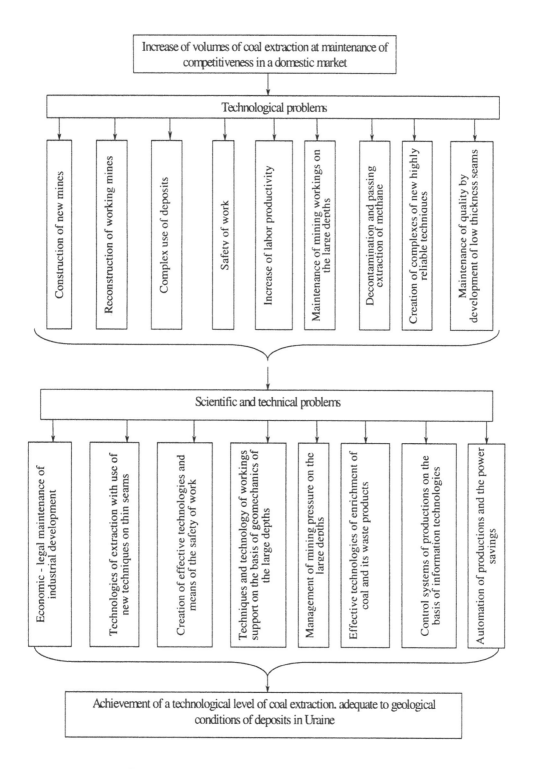

Figure 1. Scientific problems of the coal industry in Ukraine

Restriction of extraction in gassy conditions is also an issue. Preliminary degasification of coal seams and surrounding rock is required. For this purpose construction of "gas" horizons, drilling of wells from mining workings, and also from the surface with the purpose of methane extraction is necessary.

As examples of solving technological problems by means of achievement of mining science and mining engineering, we shall result the activity of the coal industry of Ukraine.

Recently positive process of increasing output from mechanized longwall is planned. According to the industry statistics, the operation of five MKD90 support sets results in an annual gain of coal mining of not less than 1,3 million t. It is equivalent to input of new mine with cost of construction 0,3–0,4 billion US$.

The new equipment has the capacity to work for 15–20 thousand hours that is equivalent to 5–6 years of operation before major overhaul. New, high efficiency shearers are proposed for coal extraction, and L800D, L100D, L1200D conveyors with increased installed power and range of transportation are proposed for transport.

Besides the better operational parameters, using of new extraction techniques raised productivity by 2–3 times, increased installed power per employee and reliability. This allows to increase the lengths of longwalls up to 250–450 m, to conduct work in thin seams, sharply reduce expenditures on labour of trailer operations in longwalls. Finally this will lead to growth of average output of such systems up to 1700–1800 tons per day.

However, the new techniques conflict with the used technologies, which do not provide safe conducting of mining works. There are serious scientific and technical problems in creating new systems of preparation and development of coal seams including construction of a network of auxiliary developments, drainage horizons providing safe conducting of mining and reliable liquidation of failures.

One of the perspective directions of mining stability developed in National Mining University is based on strenghtening of rock for counteraction to mining pressure.

The solution of this problem conducted optimization principles of interaction and interference of elements of the system "rockmass – the strengthened rock – support". The features of strengthening effect resulting from the action of anchors and forcing of hardening mixtures are determined.

Hardening causes occurrence of the indignations described by power characteristics, as in cross-section, and longitudinal sections of workings. These factors generate formation of spatial changes in the sizes of a zone around the workings and load on support. In result this reasons cause essential inconstancy of interaction of the "rockmass – the strengthened rock – support" system. This has an adverse effect on its stability.

The modern theory of interaction of the "rockmass – the strengthened rock – support" system includes two interconnected components: power characteristics of a subsystem "strengthened rock – support" and mining pressure. On one hand, the interrelation will be, that the power characteristic of a subsystem influences the formation of load, with another – spatial indignations having reflected on schedule of mining pressure and transforming it.

Criteria of optimization of system interaction are applied for the solution of this problem "rockmass – the strengthened rock – support" in spatial statement. It has allowed establishing required (in a specific geological situation) constructive – technological parameters of a "rockmass – the strengthened rock – support" subsystem. This with adaptation to character of mining pressure allows realizing effective technologies of supporting and maintaining mining workings.

Essentially new method of forecast of the mining stress, taking into account spatial periodic changes of interaction in "rockmass – the strengthened rock – support" system is developed. It has created preconditions for development of the method of tensely – deformed determination of the subsystem possessing spatial heterogeneity of geometrical, mechanical and power parameters with the purpose of definition and optimization of its power characteristic.

Increase of safety level of works in coal mining, in our opinion, can be provided by creating a rotection system in the geological environment with adequate modern techniques and technologies, the bichemical processes proceeding on great depths. For this purpose it is necessary to solve such problems:

- General degassing of coal deposits.
- Scientific, technical, psychological preparation, and also retraining of personnel.
- Creation of modern control systems for mining works with use of information technologies.
- Introduction of modern and safe techniques.
- Scientific and technical support of high-productivity technologies of coal output.
- Improvement of legal base of work safety.

As an element of safety system, on the basis of scientific development of NMU for mine-rescue works, prototypes of a mobile rescue elevating installation ASPPU-6,3 were constructed in Novokramatorsk machine-building and Donetsk experimental mechanical-repair factories. Its acceptance tests were successfully conducted in December 2003 on Kalinin mine.

After tests installation it was transferred to an operative mine-rescue group (Donetsk) for constant operation.

One of the priority directions of continuous fuel coal supply for electric power generating industry is to increase its quality up to economically expedient level. It will allow to lower the unit cost price of the electric power production and to raise the volume from extracted coal, to lower ecological damage, to raise economic efficiency of the coal industry.

However, the features of coal deposits located in the territory of Ukraine predetermine rather high cost price of extraction and large ash content. These features, and also deficiency of fuel, social and economic problems result that actual ash content changes within the 35–55%.

Further ash content increase will lead to growth of the cost price of the enriched fuel and deterioration of economic parameters of the coal industry and electric power industry.

In the case of using coal with large ash content in area of its extraction, more effective becomes modernization of the existing and development of new methods of it's burning. The given questions are expedient for considering in a context of a modern condition of power production from solid fuel.

Problem of the work cycle extraction and coal usage is production of electric energy. Its minimal cost price has to recognize the given natural properties of energy carriers and a technological level of power.

The question of substantiation of coal quality should be considered from the position of completeness of use of its energy potential, technological opportunities of its achievement and economic feasibility.

If we start with consumer properties of fuel at its optimum quality, obtaining the maximum quantity of useful heat is possible. Research executed in NMU, have shown, that the maximum quantity of useful heat is produced at ash content of the coal concentrates, corresponding parent ash content of pure coal (fig. 2).

For decades of work the coal industry near mines and processing plants has collected more than 120 million tons (and continues to collect) of coal bulks, which represent secondary fuel resource which can be used and can improve fuel and energy balance due to manufacture from waste products of composite coal fuel for household needs.

Thus, there was a problem of recycling of waste products of dressing the coal output. Ash content of such waste products reaches 60–80% at heat of combustion of 7,5–10 MJ/KG. That is caused by losses of coal fractions. For thermal processing of such waste products it is necessary to introduce new technologies and create generating capacities for their transportation.

Usage of traditional technologies of coal dressing, the most widespread on Ukraine and in the CIS, results in the raised losses of combustible weight.

The solution to this problem – to increase the degree of use of an energy potential extracted from coal is possible, on the one hand, by allocation of intermediate products with ash content up to 60–65% in existing processing plants and their burning at the place of their extraction with use of special technologies (CPS-technologies). On other hand – by application of coal dressing technologies, adapted to modern granule structure.

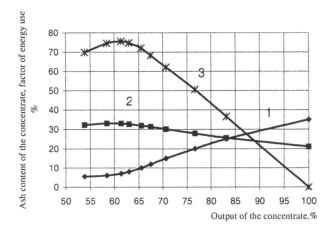

Figure 2. Parameters of energetic use of coal.
1 – ash content of the concentrate; 2 – factor of energy use of coal at only concentrate burning;
3 – criterion of technological efficiency

Application in coal dressing of rather simple gravitational devices, such as screw and cone separators, investigated in National Mining University, creation in Ukraine a concentrating complex with heavy environment hydrocyclones, the organization of own manufacture of screw separators, open new technological opportunities. Already realistic is an application of technologies of coal dressing with the use of four or even five machine classes. On the basis of long-term researches the effective technology of dressing coal with separate processing four machine classes is synthesized: >13 mm – heavy duty separators with static conditions of division, 2–13 mm – heavy duty hydrocyclones, 0,35––2 mm – screw separators with cleaning of heavy fraction and <0,35 mm – flotation with cleaning chamber product, and at ash content of this class >70% can be considered as waste products.

Lump coal waste products and brown coal represent an essential fuel resource for household consumption. For this purpose new technology of coal particle binding is developed in NMU. In its basis are the chemical processes proceeding in viscous plastic systems, forming thin dispersion particles of mineral coal which cause knitting properties lay.

Using these properties it is possible to make composite fuel which components are the mix of coal and anthracite waste products, brown coal, a various sort the waste products containing organic substances.

The briquette fuel possesses high heat content and mechanical properties, in particular, sufficient mechanical durability, water resistance, thermo stability. The layers of such fuel during burning have good gas permeability that provides full enough degree of combustion even at high ash content.

This technology is not power capacity. About 10 KWH are spent for ton of finished goods. Drying is carried out in natural conditions or with blowing of hot air during the autumn-winter period. The cost price of the limit is 5–6 dollars per ton.

Ready fuel has heat content of not less than 3500 kcal/kg, and at a piece brown coal with low ash coal reached 4500 kcal/kg.

The ready product looks like cylindrical cores with diameter of 30 mm and length up to 200 mm.

Installations using this technology are already made in small series under the order of customers.

To increase efficiency of utilization of energy potential of the extracted coal the following can be done:

– Producing in processing plants coal concentrates with smaller ash content for effective burning in existing power electric stations. This will allow receiving useful heat and essentially lower transport charges.

– Construction of power units that use CKS-technologies in coal-mining regions with significant volumes of accumulated coal waste products.

Such ideology will allow realizing the energy potential of combustible weight of coal, save fuel resources of the country, reduce the price of electric power produced on the basis of solid fuel.

Efficiency of modern mining can be achieved only with widespread use of information technologies and automation on the basis of computer systems. An essential reserve of coals output increase is reduction of idle times of technological systems, which are estimated to cause losses of extraction of 20 million t/year. Decrease in the charge and losses of electric power at all sites of mining also are essential factors increasing economic parameters. The potential power saving in the coal industry is estimated at 450 million KWh per year.

The solution of this problem in NMU is based on revealing price properties of power parameters of mining machinery and installations as hyperbolic distributions. It allows proceeding from normalized power usage to power control. It raises the level of electro uses.

Analytical maintenance and corresponding monitoring systems are necessary for realization of such approach and for automation of management by separate parts of technological process of extraction.

Objective information on operating modes of the mining equipment and systems of power supply allows to reveal and locate damages to mining - technological complexes quickly and more precisely. This considerably reduces their idle time that predetermines growth of the volumes of extraction. The technological level of operation and safety of works rises simultaneously.

The new technology is most effectively realized in local (not centralized) systems with computerised controllers and multilevel architecture of management. Presence of similar systems creates a basis for the operative solutions of problems, managements of technological processes and reductions of idle times in all parts of production.

On the basis of such approach and the developed algorithms a number of computer systems ready for introduction are created in NMU.

Progress in the coal industry is impossible without a modern economic mining science. Special interest is represented with the questions connected to attraction of investments into the coal industry.

Investment problems of coal branch NMU considers from positions of how mines are attractive to investment, instead of cost of their actives. High cost of the basic actives of mine does not testify at all to its profitability. Profitability of coal extraction depends on prospect of seams development in borders of the mine's lease area. The second major part of mine capacity support is addressing investments (investment appeal).

Effective development of mine fund provides use of the offered mechanism of the quantitative estimation (the integrated parameter) conditions of each mine (technical, economic balance) taking into account attractiveness to investments (investment appeal of mines). The volume of capital investments depends on the level of investment reliability for a gain of 1 t of output.

CONCLUSION

The developed strategy of coal industry development in Ukraine is focused on increases of coal use for power generating.

Its realization will demand increase in capacities due to development of new deposits, both by construction of new mines, and by reconstruction of working ones. Thus principles of integrated approach, correct coal use and maximum use of the existing industrial infrastructure should be observed.

At the same time, apart from creation of attractive investment conditions in the coal industry of Ukraine, serious contribution of the mining science for the solution of those scientific problems, which are put forward by geological conditions of underground extraction and modern economy is also needed.

The presented examples testify to the progress in creation of new technologies of thin coal seams development and complex processing of mining output.

They prove that scientific priorities stem from the necessity to find solution of scientific and technical problems of the coal industry at the up-to-date level.

International Mining Forum 2005, Sobczyk & Kicki (eds) © 2005 Taylor & Francis Group, London, ISBN 0415 375525

Environmental Aspects of Mine Closure – Polish Experience

Jan Palarski
Silesian Technical University of Gliwice (STU), Faculty of Mining & Geology
Institute of Geotechnology, Geophysics & Ecology of Industrial Areas. Gliwice, Poland

ABSTRACT: The results obtained from the monitoring program of mine closure and post-closure processes are presented in this paper. The objective of the article is to evaluate how mine closure and rehabilitation can be best designed in order for the mine to contribute to the long-term sustainable development of mining sites. Mine closure plan should include site closure issues as well as environmental, economic, social and employee matters.

KEYWORDS: Mine closure, sustainable development, rehabilitation, water contamination

1. INTRODUCTION

Active mine that extracts mineral deposit creates work places and supplies raw material necessary for the development of modern civilization. However, it also damages the natural environment, quite often causing irreversible changes. So question arises – can mining activity be treated as a process contributing to sustainable development? Looking directly at the definition of sustainable development it might be stated that mining cannot be sustainable because it extracts finite resources and reduces the potential for future generations. Yet, it is a narrow interpretation of the idea of sustainability. In reality, mining should contribute to the principles of sustainable development by maximizing productivity, economic and socio-cultural benefits from mining extraction that aim to improve life standards of present and future generations. Mining activity is a temporally usage of a certain area, which during the extraction process, provides profits to its owner and local community as well as stimulating to the region. Status achieved and capital earned during the mining activity, can stimulate future development of a region after the mine closure (Report 2002).

Mine extracts non-renewable resource. After some time the extraction process is stopped and finally, a mine is abandoned and closed. Mine closure may result from other factors such as economic and ecological issues. Mine closure process should be carried out following the aforementioned development principles, as well as other i.e.:

- Creation of maximum number of work places on abandoned mine assets.
- Returning the society mined-out areas adapted for the industrial and/or civil building or at least rehabilitated for the recreation purposes.
- Protection of the remaining part of the extracted deposit and accompanying minerals.
- Reduction of damage done to the environment and water by the closed mine.

In the paper the experience achieved during the mine closure process of hard coal, ore, salt and sulfur mines are presented. Special attention is given to safety rules of closing the underground workings, minimization of damage done to the environment and hazard monitoring on closed sites. It should be highlighted that a mine closure process needs to be understood as an earlier planned stage of the mine's "life cycle", not as a social or economic failure and ecological disaster of the region where the mine is located.

The paper presents basic and verified rules of the mine closure policy process as well as the author's point of view on the matter of mine closure and mineral resources economy policy in Poland.

2. PRINCIPLES OF MINE CLOSURE PROCESS

As it was stated in the introduction, a process of mineral extraction is based on an non-renewable finite resources. Moreover, extraction can be done only where a deposit is located. It may sometimes mean inconvenient conditions. Contemporary technologies enable the extraction of accessible deposits in a relatively short time. So far, mines were operating for hundreds of years. Nowadays, mines operate approximately 15 to 50 years. Mining operation may positively contribute to the sustainable development. It is possible only when a long-term impact assessment on the environment through the whole life cycle of the mine is done. The assessment should be placed in the preliminary mine closure plan. Preliminary mine closure plan ought to include both negative and positive impact forecasts of every stage of mine operation on the considered mining area. It should be updated and corrected during the development of mine operation and after the receipt of results. The plan needs to formulate rules that minimize the negative impact of extraction and mineral processing on the environment with regards to available technologies of extraction, current law in force and social interest.

The main idea of final closure plan is to elaborate technological solution of mine closure and rehabilitation of mining area on the basis of information obtained from the preliminary closure plan regarding: health safety rules, social interest of employees and inhabitants, work safety and environment protection - bearing in mind closure cost restrictions and further stable economic growth of the region.

After mine closure it is possible to improve inhabitant's life standards only when the surface of mining area is properly rehabilitated. New investors, local community's self development and creativity are needed. All these processes should be conducted with the great engagement of local authorities and mine owner.

In case of mine having no mine closure plan and where the decision of closure was made up suddenly it is impossible for the sustainable development conditions to be fulfilled. What is more, it does not have sufficient funds to sustain the process. That state of affairs may be currently observed in Polish closed collieries and other mines. In some cases, that might lead to ecological, social and economic disasters which effects will be seen for years. It is slightly easier to prevent from that kind of occurrence when the mine future production is supposed to continue for several years. In that situation, it is possible to correct hitherto existing negligence, collect necessary funds for solving social problems, rehabilitating the environment, securing the abandoned mine and above all, elaborating the mine closure plan. Polish mines should elaborate such programs since it seems to be the only way to solve the existing problems of closed mines. Legal regulation changes regarding these issues are necessary. That ought to make companies elaborate their closure plans and its updates. Frame range of underground, surface and borehole closure plans was quoted in paper (Palarski 2002).

3. MINE CLOSURE PROBLEMS

Decommissioning (physical closure of mine) includes:
- Removal of all or some part of the equipment.
- Secure, filling, sealing and adaptation workings.
- Minimization of damage done to environment and restoration of the site (i.e. water pumping, gas capture, construction of clay seals).

- Installation of monitoring equipment.
- Rehabilitation of surface mining area with waste dumps.
  These processes are most frequently conducted at the same time depending on the type of a mine, deposit and its location, range of a mining activity and objective set for it. Works are done according to the mine closure work plan that is based on the mine closure plan. Final closure plan based on the existing preliminary plan must be completed by:
  - Study projects of currently available methods and technologies used in abandonment workings.
  - Analysis of feasible rehabilitation methods in the considered area.
  - Consulting the project of area rehabilitation with all the interested parties in order to work out the best solution.
  - Sample tests and expert's reports done to choose the best available technology for mine closure and rehabilitation of the site.

After such analyses are conducted it is possible to:
- Calculate costs of possible technologies of mine closure and site rehabilitation including various methods and materials.
- Choose monitoring methods – find optimal solution for an objective that is set for the area acceptable by the local community, fulfilling the environmental protection requirements and available funds.

It is known that all closure, rehabilitation and monitoring procedures must be consistent with the current law regulations, especially with mining and geological, water, environmental protection and labour laws. The closure risks include environmental impacts, safety and health, social, legal, financial, technical and other issues.

So far, mine closure was carried out in the following mines in Poland (table 1):
- Underground mines of iron, lead and zinc, copper, barite, salt, sulfur, limestone, brown and hard coal.
- Open cast mines of different materials.
- Borehole mines of salt and sulfur.
- Oil and gas mines as well as mineral water production plants.

Table 1. Closed mines in Poland in 2003

RESOURCES	NUMBER OF CLOSED MINES	NUMBER OF MINES IN THE PROCESS OF CLOSURE
Hard coal	6	1
Lead and zinc	-	2
Copper	1	-
Salt	4	-
Sulfur (bore-hole)	2	-
Gas and oil	15	-
Other minerals	approximately 250	-

4. ECONOMIC AND ECOLOGICAL ASPECTS OF MINE CLOSURE

Observation done so far indicates that in most cases it is very difficult to eliminate completely the negative impact of mine activity on the environment. However, the natural environment adapts to existing conditions and after some time eliminates hazards. Unfortunately, not all of them can be eliminated. Some like surface damages, water contamination, spontaneous combustions, gas emissions are noticed even after a several years. It indicates that most of mine closure has been done improperly and constitutes danger for the future years (Palarski 2003).

4.1. *Mineral resources economy*

The protection of resources during mining operation and closure is fundamental matter. It is clearly seen on the example of Polish coal mining industry. In 1990 when Poland has undertaken the process of restructuring the mining industry the amount of measured resources was proved as 30 billion metric tons in developed mining areas and 35 in undeveloped. At the end of 2003 is amounted respectively about 16 billion and 28 billion metric tons. The amount of recoverable resources at that time decreased from approximately 17 billion to 7 billion metric tons. Ranging from 1990 till the end of 2003 the amount of 1,58 billion t of coal was extracted which means that 8,4 billion t of recoverable resources in the balance was lost (figure 1, 2). Yet, there is over 1,7 billion t of minable coal reserves in the seams over 2 m thick.

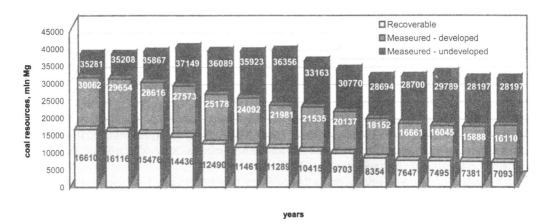

Figure 1. Polish coal resources

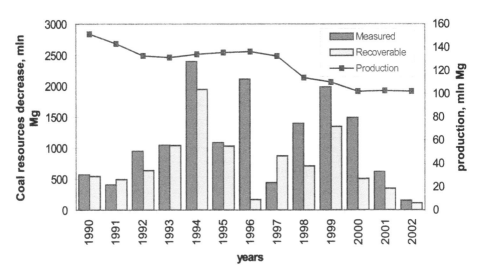

Figure 2. Reserve decrease in developed seams

36

Without developing new methods of extraction of thin seams, accessing new deposits and considering the same level of production, coal extraction in Poland will continue only for the next 10 to 15 years. The reduction of coal resources is the result of coal restructuring program strategy. Its purpose was to enhance the productivity by mine closure and abandonment of production from deposits under difficult geological conditions.

Similarly, the matter exists in ore and other raw materials mining.

4.2. *Rock mass and surface deformation*

Underground mining activity causes rock mass and surface deformation of various range and results. After mine closure, voids and ungrouted gobs may still cause surface subsidence, sinkholes and other damages. Long term processes of mine flooding, coal pillars failure or underground fire of coal or wood could put the surface into sudden danger. That may happen even after several years from mine abandonment.

Additionally, in the process of mine closure, harms caused to the environment while active mining need to be reduced. Such problem frequently appears at the Upper Silesia Coal Basin. However, abandoned mine areas where salt, brown coal, zinc and lead, copper and sulfur (bore-hole mining method) were mined, deal with similar difficulties.

Measurements of surface subsidence and deformation on the abandoned mine area indicate that the deformation process ceases almost completely after 3 to 5 years from the cessation of production (region of Dabrowa Gornicza, Sosnowiec and surrounding area). At the Lower Silesia Coal Basin the time span record was 4 to 6 years. For example, at the "P-K" coal mine during the 5 year monitoring period, the average horizontal strains change measured close to the main shafts were approximately -0,5 mm/m/year. Strains measured close to the local road were as slight as -2,7 mm/m in the period of 3 years. However, significant strains change in the years: 1995 – -2,5 mm/m, 1996 – -2,9 mm/m, 1997 – -3,6 mm/m, 1998 – -3,5 mm/m, 1999 – -3,7 mm/m and in 2000 – -3,8 mm/m were recorded on the measuring line situated close to the "J" shaft.

Locally, some subsidence during the mine flooding was observed. The sudden movement of underground water could explain it after the pillar or dam failure. The result is an underground collapse of rock mass.

Special attention has to be paid to the shallow underground workings (up to 60 m deep) causing surface damages (crown hole, chimney caving, plug subsidence, etc.). At the Upper Silesia Coal Basin where mining was carried out at the depth of at least 100 m continuous subsidence were not recorded, excluding fault zones.

Sudden outflow of water with sand or clay from the water-bearing bed to the workings may cause rock mass deformation (i.e. surface discontinuous deformation caused by the water outflow to the one of the salt mine adits).

4.3. *Seismic activity of rock mass*

The analysis of seismic activity after the mine closure is very difficult due to the abandonment of seismic monitoring. Usually, after mine closure process seismic station monitoring is stopped. However, in some mines, post-closure seismic activity was still observed. General conclusion to be drawn is that cessation of mining decreases the energy and number of tremors (recorded tremors energies were in the range of 10^2–10^4 J). In several cases, even a few years after the mine closure seismic activity is still measured and it depends mostly on the properties of rock mass, depth of extraction and level of water in the flooded mine. Water affects the slide resistance of fractured rock mass and change in the stress distribution. Surveys of seismic activity in the closed mines are quite complex because of the neighbouring mines activities. The amount of tremors and induced energy is difficult to attribute to a particular mining activity. However, it might be stated with great certainty that seismic activity does not stop immediately after the mine closure.

4.4. Hydrological changes and chemical composition of mine water

During the mining operation rock mass is drained and natural water conditions are often indivertibly changed. The changes are frequently much wider then mining area itself. Stopping of water pumping after the mine closure causes mine workings flooding and gradual uprising of groundwater table to the natural level. That may constitute danger to the surface infrastructure. The problem of surface flooding has been observed in almost all type of mining operation – underground, open pit and borehole. There are many examples of such situation occurrence, especially on the areas where water pumping was stopped. If the underground workings are isolated from the surface waters and overlaying aquifer by the impermeable bed, even during the mining cperation surface subsidence and sinkholes may appear.

Not only flooding of the pit but also by surrounded drainage area loes restoration of natural water conditions in open cast mines. For example, in the "P" sulfur mine the impact on the surrounded area was more than 40 km^2. Abandonment of the iron mines "W" and "S" in the Czestochowa-Klobuck region lead to change of groundwater flow, flooding of surface buildings, local depressions, outflow of water from monitoring wells and changes in chemical composition of water.

The closure of mines can result in the accumulation of mine waters, which can pose a threat of vast quantities water inrush into workings of surrounding mines. That is why in closed mines the water needs to be pumped. Various strategically located pumping stations often protect the operating mines.

Another significant problem connected with the mine closure caused by flooding is the change in chemical composition of groundwater and its impact on the surface water. Any closure of driven workings in the field of mining operation not only increases the water inflow to other parts of the mine but also changes the mineralization of water and its reaction (figure 3). As it is presented on the chart, there are no significant changes of water quality in the hard coal mines. The only exception is water flowing through the backfilled areas (fly ash and mineral processing waste as fill materials) and through rocks with high pyrite content.

Some serious changes in water chemical composition were noticed in the flooded iron mines. Concentration of SO_4^{2-}, Mn^+ and Fe^+ ions in the surface water exceeded the standards respectively: 7-times, 100-times and 500-times. It was the result of oxidation, hydrolysis and lixiviation of ions from the rocks surrounding flooded mine workings and their seepage in the groundwater up to the surface.

Notable changes in water chemical composition were also noticed during and after the sulfur open pit and borehole mine closure. Chemical degradation of subsurface water is caused by the contact with sulfur deposit and some factors such as:
– Washing out the sulfur from the boundary fields and dumps by the rainfall.
– Surface dewatering of boundary fields with high acidity soil (infiltration water).
– Leaching of Al^+, Fe^+, Cr^+, Ni^+, Zn^+, Mg^+, P^+, K^+, SO_4^{2-} and Cl^- ions.

Water pumped from the "G" sulfur mine had an average content of sulfates – 1491 mg/dm^3, chlorides – 226 mg/dm^3 and the average pH was 3,97 (minimum 1,8). Because of that high acidity water has to be neutralized, for instance by adding of calcium, before it is diverted.

High contamination of groundwater and surface water is also present during and after the lead and zinc mine closure. The example is a 130 km^2 area of a groundwater basin "Bytom", located in the highly porous and fractured ore-bearing dolomite in the lead and zinc industry zone. Water strongly contaminated by dissolved iron, sulfate and carbonate ions (Zn^+, Pb^+, Cu^+, Fe^+, Cd^+, Sb^+, As^+, $CaSO_4$, $CaCO_3$) may constitute great danger for the surface water.

Moreover, in the borehole sulfur, oil and gas mines the danger of water eruptions and sudden emissions of sulphuretted hydrogen and methane exist. It is caused by the rock mass (deposit) relaxation and outflow through fractures and leaks around the wells. To eliminate this danger special preventive actions during the mine closure are undertaken. Above all, permanent and dense capping needs to be done.

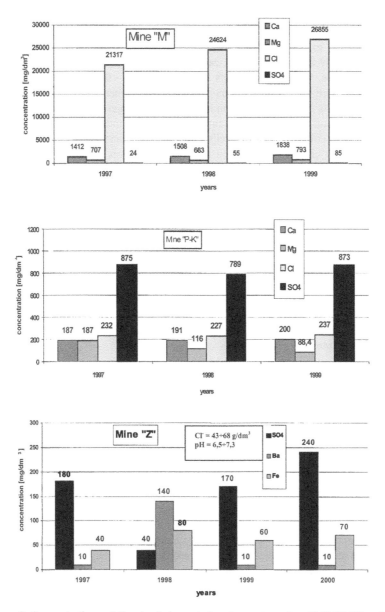

Figure 3. Concentrations of dissolved elements in mine water; mines "M", "P-K" and "Z"

4.5. *Problems connected with soil contamination*

During the mine closure process, ruined soil must be rehabilitated. Mining has a negative impact on the soils situated on the mining area. It is caused by the drilling, road building, installation of technological sites, depositing of mineral and dumping of waste, etc. That may lead to water pH decrease (1,2–2,2) as well as contamination by heavy metals, or in some mines, radioactive elements. The main sources of contamination are water flown and dust blown from the waste dumps, drainage from the tailings ponds and waste water plants, drilling mud and all kinds of leakage caused by the damages and eruptions.

4.6. *Gas emission danger on mining areas*

The process of mineral extraction is followed by the emission and migration of gases trapped in a deposit. Most frequently these are methane, carbon dioxide or sulphuretted hydrogen. Gas emissions are dangerous for miners, region's inhabitants and environment – that is not only during the mining operation and mine closure but also many years after these. Detainment of underground mine ventilation and water pumping results in gas accumulation in workings, fractures, apertures and pores. The upraise of groundwater level causes gas displacement up the surface and its emission to atmosphere. As a consequence, in closed collieries drainage wells and surface methane capture stations are installed. The captured gas is used for electricity generation and heat production. Methane drainage in one of the closed colliery is presented in the figure 4. Figure 4 shows significant amounts of methane that can be captured. Certainly, it is a good source of energy and it does not enhance the greenhouse effect.

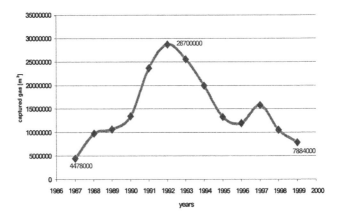

Figure 4. Methane drainage in mine "M" during 1987–1999.
Drainage in 2004: 600–1000 m³/day

5. CONCLUSIONS

1. Preliminary mine closure plan should be placed in the mine design project and updated during the development of mine operation. Preliminary mine closure plan ought to include both negative and positive impact forecasts of every stage of mine operation on the considered mining area. It should be updated and corrected during the development of mine operation and after receipt of results. The plan needs to formulate rules that minimize the negative impact of extraction and mineral processing on the environment with regard to available technologies of extraction, current law in force and social interest.
2. The main idea of a final closure plan is to elaborate technological solution of mine closure and rehabilitation of mining area on the basis of information obtained from the preliminary closure regarding: health safety rules, social interest of employees and region's inhabitants, work safety and environment protection – bearing in mind abandonment cost restrictions and further stable economic growth of the region.
3. Mining has a negative impact on the natural environment. Thus, in the process of mine closure remaining resources ought to be protected and hazards that may appear as a result of mining operation need to be removed. Rehabilitation of the mining area and surrounding site should lead to environment safety and future growth of the region.

REFERENCES

Palarski J.: Ryzyko dla środowiska w procesie likwidacji zakładów górniczych. Bezpieczeństwo Pracy i Ochrona Środowiska w Górnictwie 2002.

Palarski J.: Likwidacja kopalń a zagrożenia dla środowiska – Materiały zebrane z pomiarów w likwidowanych kopalniach. Instytut Geotechnologii, Geofizyki Górniczej i Ekologii Terenów Przemysłowych Politechniki Śląskiej w Gliwicach 2003 (materiały niepublikowane).

Report of the MMSD Project. Breaking New Ground. Earthscan Publications Ltd. London and Sterling 2002.

International Mining Forum 2005, Sobczyk & Kicki (eds) © 2005 Taylor & Francis Group, London, ISBN 0415 375525

Approaches to Systematic Mine Closure and Mine Site Rehabilitation in the Baia Mare Mining District, Romania

Nicolae Băncilă-Afrim
North University of Baia Mare. Baia Mare, Romania

Christian Buhrow
TU Bergakademie Freiberg. Freiberg, Germany

Dorel Guşat
TU Bergakademie Freiberg. Freiberg, Germany

Adalbert Kruk
"Vamera" S.R.L. Baia Mare. Baia Mare, Romania

Martin Peterfi
North University of Baia Mare. Baia Mare, Romania

1. PRESENTATION OF BAIA MARE MINING REGION

The mining region of Baia Mare is situated in the North-West of Romania. Its ore deposits are linked to the Gutâi Mountains metallogenetic alignment.

The mineralization is represented by the epithermal vein systems featuring predominantly polymetallic characteristic ± Au and Ag. Most of the deposits occur in vein form, their number exceeding 500 veins, and in many cases they are grouped in veins genetically bonded together.

The geometrical parameters of the veins are variable, their length between 300 meters and 5 kilometres (main vein of Baia Sprie), and thicknesses between some centimetres and 30–40 meters, some few meters mainly.

In depth, the veins extend to 200–400 meters, in some special cases down to 700–800 meters (Şuior Mine and Baia Sprie Mine).

The veins mineralization is irregular; the main component variation is between 40 to 100%. The vertical distribution of useful elements is regular; gold and silver in upper zone, lead and zinc in the median zone and copper at the bottom.

The ore deposits are usually embedded in igneous rocks (quartziferous andesite, augitic andesite ± hornblende, quartziferous andesite and microdiorite with augite and hornblende ± biotite). The contact between veins and host rocks is usually evident, but rarely presents an unclear delimitation.

The host rocks close to the veins have severe alterations (silicification, adularization, etc.), thicknesses of these zones vary from some tens of centimetres to tens of meters.

2. OGANIZATION OF MINES AND OF THEIR CLOSURE

From organization point of view all mining units, denominated in different historical periods, as mine or exploitation, are subordinated to The National Company of Precious and Base Metals – "REMIN" S.A – Baia Mare. A mine is divided into so called operating sectors. The ore deposit belonging to a mine is separated into a number of Mining Perimeters, which can comprise one or more veins, according to promotion class of reserves by the geological research organization and according to the confirmation by The National Resources Agency.

Mine closure procedure is regulated by Governmental Decision, which stipulates which Mining Perimeters are nominated to be closed. So, there can be situations when an entire mine is closed, or only some of its perimeters. In case of partial mine closure, it is necessary to be tailored and executed in such a manner that the activity of the running perimeters progress totally securely. These works, in many cases, ceased.

The Governmental Decisions concerning cessation of prospecting and exploring activities are released separately for every involved perimeter, which can include many specific works either on surface or underground, as pits, trenches, shafts, galleries, raises and waste dumps.

After 1990, many private activities began in the Baia Mare's mining district concerning especially quarrying and/or processing of rock (construction and ornamental rock), the owner being responsible for closure and necessary site rehabilitation works, at ceasing the activity.

3. THE BAIA MARE'S MINING PERIMETERS CLOSURE REQUIREMENTS

The reasons, which lead to closure of mining perimeters in Baia Mare mining district, are manifold, of which we mention the followings:
– Low economic potential of the majority of the mined mineral deposits.
– Lack of financial resources.
– High production costs.
– Gradual reduction, to total ceasing, of state subsidy (2007) paid to nonferrous mining industry.
– High degree of the environmental pollution.

Because of the necessity to cease work in some mining perimeters, N.C. "REMIN" S.A. Baia Mare has elaborated a restructuring schedule for the period of 1999–2006.

This program comprises closure of 20 mining perimeters and of two mineral processing plants.

In some perimeters designated for closure the activity was ceased and since 1990–1991 they were being placed on care and maintenance.

The situation of waste disposal (when scheduling the closure) was as follows:
– The total of 55 waste dumps, of which four are in use, include a volume of about 2.4 million cubic meters of waste rock, which cover more than 25 hectares.
– The 17 tailings ponds, of which four are in use and 13 are placed on care and maintenance, two of them being emergency ponds, cover in total 335 hectares, comprising 77 million tonnes. The two emergency ponds have an additional capacity of 24 tonnes.

The tailings ponds placed on care and maintenance are included in closure documentations (except the two emergency ponds).

The geological exploration working sites in Baia Mare mining district have been closed.

4. MINE CLOSURE METHODS

Closure of mining perimeters is regulated by Governmental Directives, controlling the mining activity, environmental protection and establishing/abolishing of industrial constructions and/or civil ones.

Based on these directives, General Management of Industrial Ministry has compiled a *Manual of Mining Closure* in 2001.

This manual has three distinct parts, namely:

Part I: General considerations.

Part II: Normative framework concerning mine closure.

Part III: Manual of methods concerning mine closure.

The main phases of mining perimeter closure methods are:

a. Elaboration of an activity cessation program (done by operator/owner).
b. The program assessment (done by the Project Implementation Unit – the Directorate for Capacity Conversion and Environmental Protection in the Mining Sector (PIU-DCCEPMS) – Ministry of Industry and Resources).
c. Offer to concession (done by the National Resources Agency).
d. Approval of closure (done by the Government).
e. Cessation of activity (done by operator/owner).
f. Placement on care and maintenance of the works according to documentation elaborated by operator/owner.
g. Closure project elaboration (done by operator/design institution).
h. Closure works accomplishment.
i. Landscape rehabilitation.
j. Post closure monitoring (done by operator/ design institution).

5. PROBLEMS OF BAIA MARE MINING DISTRICT'S CLOSURE OPERATIONS

– Because the auriferous ore deposits in Baia Mare basin were mined from ancient times, and parts of the old mining workings are not known and are not plotted on mine maps, these unknown works can influence negatively the host rock stability in the surface vicinity of these sites.

– Because of the nature of Baia Mare's ore deposits, the mine waters (both infiltrating and coming from underground creeks) are highly acidic with pH value of 1.5 to 5.0.

– Mining of the upper part of the auriferous deposits extends near the surface (pillars of 5–10 m). The created openings were backfilled with so called "low grade ore" or "refractory" ore. Because of perfecting the processing methods of the refractory ores, these backfill materials, together with the ore embedded in the protection pillars of the old areas were later mined out, leading to impossibility to control the surrounding rocks. Introducing and expanding the application of sublevel caving methods (80's) led to intensification of rock movement in the upper part of the ore deposits and to made it impossible to backfill the created gaps. As the result, a multitude of surface cracks and fractures were formed, facilitating the ingress of pluvial water underground.

– The sublevel caving method generates a lot of ore loses; abandoned *in situ*, simultaneously with the jointed ore of the protection pillars, represent potential pollution sources.

– Many waste deposits generated by geological exploration are in remote locations covering large areas of the surface depending on their volume. In several cases, these rock dumps contain useful elements, representing permanent polluting sources.

– Some tailing ponds contains high grade of useful elements, which can be recovered. In Baia Mare tailings from one pond were already reprocessed for their gold content.

6. PRESENTATION OF THE CLOSURE WORKS ACCOMPLISHED IN BAIA MARE MINING DISTRICT

According to the *Manual of Mining Closure* the necessary closure works for Baia Mare mining basin are to solve the following problems:
– Recovery of technical equipment and materials.
– Closure of the horizontal, inclined and vertical accesses to the mining workings.
– Dismantling of the infrastructure.
– Rehabilitation of the affected by the mining activity surface (waste dumps, tailings ponds, watercourses, access roads).
– Accomplishments in Baia Mare mining basin.

A. *Recovery of the mining equipment and machinery* was done for two reasons:
– Re-use in other active perimeters.
– Use as recycling material.

The equipment and machinery of shaft hoists, ventilation stations and transformer stations underground and on surface were recovered. The compressor stations and the workshops were dismantled soon after activities were ceased (during care and maintenance). Any specialist equipment used underground was redistributed to other active perimeters after the cessation of mining.

Equipment and metal structures recovery for recycling included the metal components of shafts, metal bins, platforms and other metal structures specific to surface activities. Underground, the recovery was limited to mine-cars and railways (on short sections especially in old galleries).

The operator did dismantling of the closed geological exploration sites, soon after the activity cessation.

The recovery of the potential polluting materials from underground (as timber and metal), was sporadic, in many cases impossible, because of the condition of the mining workings.

B. *Closure of stopes and development works* can be summed up to recovery and removal of the mined ore by the operator. Actually, closure of stopes consists in aborting them.

The closure of the Țibleș-Tomnatic perimeter was done by flooding the underground mining workings (the access ways and the adits in the upper section of the mine were sealed with walls), the water being raised for about 100 meters, and directed to a treatment station.

Considering that there are valleys situated lower than some of the mining levels (difference of 20–40 meters), and that there were some springs on the slopes before the mining activity started, it looks possible that in this case the chosen closure method is inadequate. If the sprigs reactivate the mine water output increases, leading to uncontrolled acid water burst, and pollution of the environment.

C. *The closure of access ways* is represented by sealing adits and hoisting and ventilation shafts, as well as main and secondary ventilation raises.

The adits in the first section from the entrance usually are lined by reinforced concrete, masonry or supported with metal supports. The old adits are supported with timber or are not supported at all.

The adit closure, according to regulation is accomplished by two sealing reinforced concrete or masonry walls, one being erected at the entrance and the second one at minimum 10 meters from first wall. The section between them is backfilled (see details in figure 1).

Although the technical recommendation stipulates that the length of the backfilled section in the gallery have is to be extended until the overburden rock thickness above the adit reaches 50 meters (see figure 1), the backfill extension never exceeds 20 meters (Tyuzoşa perimeter – Nistru Mine).

The problems encountered during the old adits closure vs. recommended method are presented in table 1.

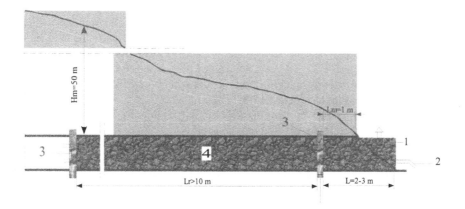

Figure 1. Recommended method for adit closure.
Hm – the overburden rock thickness; Lr – the length of backfilled section.
1 – vent pipe of 50 mm diameter; 2 – water drainage pipe of minimum 100 mm diameter;
3 – reinforced concrete or masonry wall of 30–50 mm thick; 4 – backfill;
5 – embedment alongside the perimeter of wall 50 cm deep

Table 1. The problems encountered during the old adits closure vs. recommended method

No.	THE CONDITION OF ADIT	CLOSURE METHOD	LOCATION
1.	Collapsed entrance for an unknown length	Aborting the adit	Anton perimeter Nistru Mine
		Backfilling short sections	Aluniş perimeter Ilba Mine
2.	Difficult access to the adit level	Sealing wall erection	Venera perimeter Ilba Mine
3.	Impossible to backfill along the specified minimum 10 meters from the entrance	Erection of only one wall at the portal.	Anton perimeter Nistru Mine

Regardless the number of walls, mine water drainage and evacuation of accumulated gases must be assured by installing pipes through both walls.

The method of shaft closure is illustrated in figure 2.

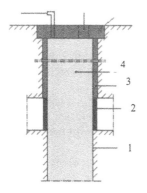

Figure 2. Shaft closure method.
1 – shaft pipe; 2 – sealing walls at plat level; 3 – the shaft lining;
4 – backfill; 5 – concrete cap; 6 – vent pipe to release gases;
7 – inspection opening and cover

The shafts are backfilled entirely. The used backfill material must not contain fractions above 250 mm and in the uppermost 50 meters of the shaft the diameter of particles is to be maximum 100 mm. The backfill material density must be higher than 1.3×10^3 Kg/m^3.

A problem with shaft closures is that shafts are divided by partitions between hoisting compartments and passageways and the platforms in them. These are not removed and backfill scarcely is able to fill the passageway, which with time leads to major fill settlement.

Rise closure is accomplished by their backfilling and sealing on the surface with reinforced concrete caps, each incorporating a gas evacuation pipe.

Problems were met with closures of secondary ventilation ways, where the access to their collars, situated in remote mountainous or woody locations, was very difficult. There was lack of water, power and the necessary filling and building materials. Some of the old raises have been collapsed at the collar. In these cases the closure of the raises have been accomplished by sealing them with concrete caps.

Closures of access ways are done using only fireproof and non-polluting materials. Every closed mining working is to be marked by a warning plate, indicating the name and the coordinates of it.

D. *Waste rock deposits closure*

From this category of repository, only waste rock dumps were subjected to closure, tailings ponds being placed on care and maintenance due to diligence of N.C. "REMIN" S.A. Baia Mare.

Closure and site rehabilitation works comprised construction of benches, stabilizing fences, and guard trenches. At every bench, platform levelling and ravine filling was constructed, in addition rehabilitation of sites by planting vegetation was done. Fuel wastes and pollutant materials were collected and deposited in specially developed places (see figure 3).

Generally, waste dump slope inclination exceed the angle perilous to vegetation growth (26–40 degrees). The washing effect of pluvial water was high with no other stabilization facilities developed along the slopes.

Special gabions or abutments were not properly designed. Waste dumps apparently showed stability.

We consider that the total stabilization works against the pluvial eroding are insufficient considering that rainfall exceeds 50 l/m^2 during the rainy season.

The special works engineered to prevent water and air ingress within the deposited rock mass (as insulating and sealing works) are insufficient for the general stability of waste dumps, especially in cases where the rock contains useful elements, facilitating the development of acidic environment and rock weathering.

In the case of wastes generated by geological exploration activities the situation is more worrying, because it is not possible to carry out re-profiling and essential stabilization works in the upper sections of the dumps. Often, the toe of a waste dump extends down in the valley and blocks the natural drainage. In remote areas waste dumps are stabilizing and revegetating naturally.

E. *Dismantling of the infrastructure and ground clearing* of closed mining sites in Baia Mare basin proceeds in the following steps:

- After fulfilling the formalities concerning the closure of a mining perimeter, local authorities are consulted regarding the end use of the infrastructure.
- Demolition and dismantling of those constructions, which haven't, any further use.
- Decontamination of the "vacant land" by removing the contaminated soil and disposal of it on special dumps.
- Land reclamation. The materials coming from dismantling/demolition, which cannot be recycled, are delivered to underground backfilling if they are suitable. The contaminated materials are deposited as previously mentioned. Waste rock and soil contaminated with fuel, lubricants, metals and chemical agents are also deposited in the same way.
- Rehabilitation of the land is accomplished by spreading topsoil on the whole surface and planting grass and trees.

F. *Surface stabilization* at many closed mining sites represents a challenging issue. Generally, backfilling of voids resulting from mining is done in the case of adits, shafts, and rises. Only in one case the pits and cracks generated by mining activity were filled (Anton II Perimeter- Nistru Mine). Some closed mining sites were delimited as risk zones. In a few cases the delimitation wasn't based on a preliminary study of the site-specific situations, probably because the object was being situated in a remote location.

As a consequence of the utilized mining method (sublevel caving), coarse backfill material and mining out of ore pillars, gaps migrated upward generating subsidences of the surface extending far beyond the mining perimeters and the outlined risk zones. (Ilba Mine; Tyuzoşa-Nistru Mine; and Socea perimeters).

Although the perimeters affected by subsidence are remote from human settlements, there is always a danger of people or animals accessing the industrial sites.

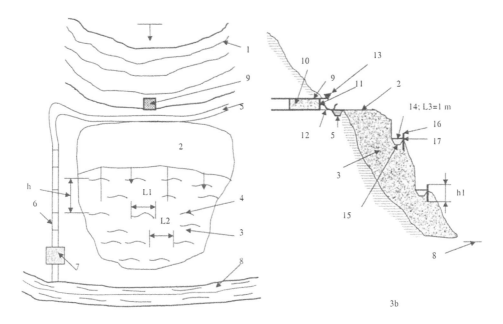

Figure 3. Waste rock dump stabilization.
1 – contour lines; 2 – platform with reversed slope; 3 – waste rock dump; 4 – stabilizing fences of
L1 = 50 meter length at L2 = 4 meter distance; 5 – guard trench; 6 – cascade canal; 7 – desander; 8 – creek;
9 – adit; 10 – backfill; 11 – wall; 12 – mine water drainage pipe; 13 – gas escape pipe; 14 – bench of
L3 = 1 meter width, reversed slope of 4 degrees; 15 – pit for vegetation excavated at 2–4 meter,
0.8 × 0.8 × 0.4 meter, pit slope inclination of 2/1; 16 – stake h1 = 1.1 meter long (0.4 meter above the main
level), 0.08 m diameter, placed every 1 meter; 17 – wattle of 0.05 meter on a height of 0.4 meter;
h – the distance between benches

G. *Reclaiming underground and surface water works* represents an important part of mine closure practiced in Baia Mare's mining basin.

Mine waters have two aspects: the chemical composition and erosion effect. Mine waters of the closed perimeters are mostly acidic and inflows vary from a few litres to tens of litres per minutes, but there are adits where water is missing. Water removal is performed as shown in figure 1. The removed minewaters are directed toward guard canals protecting dump slopes from water flow.

Water collected in guard canals is discharged into creeks, rivers, or streams only when the levels of pollutants it contains are low. Otherwise it is directed to existing treatment stations.

The benches' platforms are constructed with reverse slopes whereby surface water is collected by guard canals and cascade canals, situated alongside the benches' slopes.

In cases when water is discharged directly to river streams, a desander retaining the suspended solids is installed prior to the discharge place.

Water treatment stations are erected only when water volumes or water pollutant content are high, and they are economically justified.

When useful elements are present in waste dump rock, water ingress will generate acidic environment, leading to heavy metal contamination of the water spills. The output and the level of useful elements content in the waters are unknown, so the treatment of them is not solved.

To attenuate pluvial water erosion effect in closed mining perimeters the following works were performed:

- Pluvial water collecting systems. The canals being lined by slabs or paved (see figure 3a).
- Water flow breaking benches (see figure 3b).
- Erosion fences along the slopes (see figure 3a and b).
- Ravine filling and weir installation.
- River stream landscaping within the closed perimeters.
- Planting of grass and trees on stabilizing benches.

7. IN LIEU OF CONCLUSIONS

When evaluating the already closed mining perimeters in Baia Mare basin after about four years, it emerges that generally the accomplished works matches the design, but in a few cases, some improvements to meet the site-specific features should be necessary.

In case of mining perimeter closures the following problems have been solved:

a. Prevent human and animal access to the closed mining workings.
b. Subsidence stabilization.
c. Rehabilitation of closed mining sites.

By closing and filling the entrances to the underground access to the site is prevented, but in subsidence zone access prevention cannot be achieved.

Photo 1

The presence of subsidences, cracks, pits, irregularities, and craters outside the risk zone assist an uncontrolled ingress of waters supplying the underground water output, which could lead to compromising the treatment stations.

Waste dumps were not problematic with regard to their overall stability. But, in cases of steep slope sides and small number of break fences, the erosion effect of torrential rains is more accentuated and generates a continuous slope movement. Granular composition of waste dumps contributes greatly to this effect, too (see photos 1, 2 and 3).

Photo 2

Photo 3

Surface and platforms of the benches were well constructed, but the dump slopes stability is not satisfactory.

The techniques selected for revegetation are not adequate for steep dump slopes and the preservation technology ought to be changed, supported by the situation presented in photo 3.

The selected young plants are not appropriate for the acidic rock waste content of dumps, and because of the top soil fill being washed out from the prepared pits, the plant roots cannot strengthen and the plants will not survive.

The mining site decontamination was done acceptably; the water treatment stations process mine waters to permissible parameters. Where the deposited rock contains useful elements, it is important to monitor the parameters of water flowing in and out, or elaborating some sealing techniques.

The use of passive water decontamination is also a feasible technique in case of low water output.

Possibility of processing waste materials where appropriate has not been taken into consideration in mine closure programs as well.

Mine closure projects omitted the opportunity of using the underground voids or spaces where it would be possible, either for mining museums, waste disposal sites or for other purposes. There are many worldwide examples of similar applications.

We believe that long-term cooperation on this subject with the Technical University of Bergakademie Freiberg, Germany and other research organizations should be beneficial.

To meet this suggestion, a project proposal for a selected mine site closure has been elaborated together with the Technical University of Bergakademie Freiberg.

REFERENCES

Chindriş G., Kruk A.: Alternative Ways for Mine Closure. International Symposium – Technical University of Petroşani 1999.
Fodor D.: Mining and Mineral Processing Influence Upon the Environmental Features. Revista Minelor, No. 12/2001.
Kruk A., Chindriş G.: Some Technical Aspects of the Mining Perimeter Closure. International Symposium, Deva 2000.
Sasson M., Armstrong W., Crivet T.: Mine Closure in Romania. 4th World Mining Environment Congress. Romania 2001.
Mine Closure Manual. Revista Minelor, No. 10–11/2001.
The Aluniş, Faţa Mare, Valea Colbului Mining Perimeters Closure. Ilba Mine. Technical Project, ICPM S.A. Baia Mare 1999.
Tyuzoşa Mining Perimeter Closure. Nistru Mine 1999. Technical Project, ICPM S.A. Baia Mare 1999.

International Mining Forum 2005, Sobczyk & Kicki (eds) © 2005 Taylor & Francis Group, London, ISBN 0415 375525

Separation of Roof Rock Observed in Headings Under Development

Tadeusz Majcherczyk
AGH – University of Science and Technology. Cracow, Poland

Piotr Małkowski
AGH – University of Science and Technology. Cracow, Poland

Zbigniew Niedbalski
AGH – University of Science and Technology. Cracow, Poland

ABSTRACT: The paper presents a methodology of measurements and results of selected evaluative techniques of roof rock separation in headings driven in coal-mines. The results quoted were obtained by measurements done specifically for the purpose, whose aim was to determine movement of the rock mass around the headings. The extent of separation depends on the type and physical qualities of roof layers, mining conditions, and type of support applied in the heading. Due to changing mining and geological conditions around the controlled heading, the evaluation of separation should be carried out in its longest section. This allows for an appropriate evaluation of the fracture zone around the tunnel and then may become a base for the determination of load exerted by the failed roof rock on the installed support.

1. INTRODUCTION

The extent of the fracture zone around a heading is one of the factors affecting the value of support's load. An estimation of the range of the fracture zone is most often carried out using relationships known from the literature dealing with rock engineering, for instance, the theories of Protodiakonow, Sałustowicz, Cymbarewicz, Bieniawski etc. (Bieniawski 1989), (Chudek 2002), (Kłeczek 1994). However, for particular mining and geological conditions it is important to verify the obtained results of the fracture zone range by means of further measurements and to use values obtained in the site for the sake of support designs that follow (Majcherczyk, Niedbalski 2003). Such strategy will not only allow improving safety in headings being developed, but also to more effectively apply support parameters.

Nowadays the possibilities of carrying out measurements of this kind in coal mining conditions are wide, which is proven by a great variety of instrumentation applied. The following devices are commonly used at present: aerometric probe, extensometric probe (Korzeniowski 1998), endoscope (Majcherczyk, Małkowski 2002), indicatory or automatic telltales.

The paper presents measurement results of roof rock separation obtained in headings driven in coal mines. The results quoted come from these headings, in which telltales, extensometric probes and endoscopes were used, complemented by geodesic measurements.

2. INSTRUMENTATION USED

2.1. *Measurement of separation*

In order to determine separation, simple cable telltales (fig. 1a) and extensometric probes (fig. 1b) were applied. Cable telltales are used to determine separation in selected blocks of roof rocks, whereas the extensometric probes make it possible to determine displacements in several or even dozens of rock blocks on the basis of the change of magnetic field occurring around particular sensors (magnetic bolts). Readings are taken automatically using a coal-mine teletechnological network or after plugging a reading device in. The determination of separation with the use of cable telltales is based on changing position of a measurement tube in relation to the initial measurement.

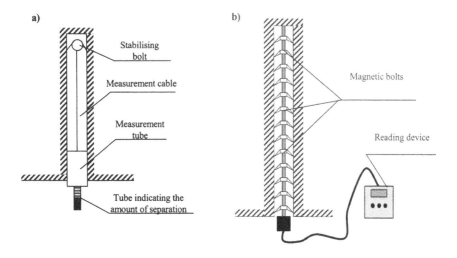

Figure 1. The devices used to measure separation:
a) cable telltale, b) multi-level extensometric probe

2.2. *Measurement of extent of fracture zone*

The way of recording changes of separations in immediate rock mass is different in the case of using borehole endoscope (figure 2), as continuous observation of walls of a borehole (core) is carried out here. It allows for the determination of the extent of separations and their precise position, the character of fissures, their inclination and the range of the fracture zone. The fact that observations can be carried out in boreholes drilled from a heading inclined at any angle in relation to the vertical axis is the biggest advantage of using an endoscope.

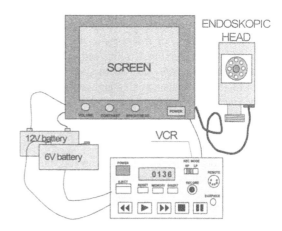

Figure 2. A schematic diagram of a borehole endoscope

2.3. Measurement of roof convergence

Measurements of roof convergence were carried out using geodesic methods on the basis of three datum points (short bolts) installed in the middle section of the roof and in the sidewalls at the height of approximately 1.5–1.7 m. The subsidence of the roof was determined on the basis of the change of the measured distance between the datum point installed in the roof and the horizontal plane between two datum points installed in the sidewalls. Such a measurement of roof layers subsidence is a relative measurement, since the datum points in the sidewalls, in relation to which the measurement was taken, are also subject to displacement.

3. RESULTS OF MEASUREMENTS

3.1. Comparison of results obtained from measurements using telltales and endoscope

The results presented (Niedbalski 2003) were obtained in the Izn inclined drift situated at the depth of approximately 900 m. The analysed heading was driven in a coal seam with the thickness of 2.5–2.8 m. In the roof of the inclined drift there were mainly clay slates and arenaceous shales with slight interlayers of sandstone. The measurements were carried out during approx. 15 months in a 100-metre length of the heading. The support at the site consisted of ŁP9/V29 steel sets comprising cross-bar arches reinforced by two pairs of bolts, installed at 1.2 m spacings. Four stations for measuring low and high separation were installed in the examined section at a distance of 25 m from the Izn inclined drift as well as one core borehole was drilled for the sake of an endoscopic measurement.

Figure 3 presents the results of low and high separation measurements. From the analysis of the measurements it may be assumed that the changes occurred mainly in the initial period of measurements, i.e. in the period of four to six months. The values of separation measured in this period were between 18 and 20 mm. Further measurements did not reveal any serious changes. The ultimate values of layer separation of roof rocks ranged between -11 mm (RN3) and -20 mm (RW4). There were only insignificant differences (maximum 4 mm) between high and low separations, which proved that the failure of rocks occurred mainly in the immediate roof up to the depth of 2.5 m. The most significant separation in the rock block from 2.5 m to 5.5 m is only 3 mm.

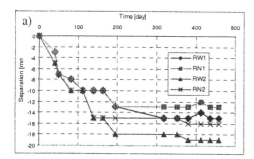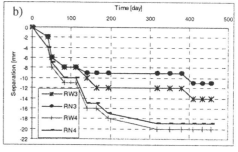

Figure 3. Change of separation in time in the Izn inclined drift:
a) telltales No. 1 and No. 2, b) telltales No. 3 and No. 4

Figure 4 presents the results of endoscopic observations. The broken line symbolises fracture, whereas the full line represents fissure. On the basis of an analysis of the results, it may be concluded that the extent of the fracture zone between the first and the last measurement changed insignificantly from 0.45 m to 0.90 m.

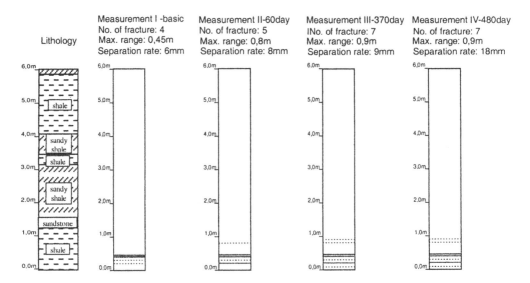

Figure 4. Results of endoscopic observations in consecutive measurement series
for the hole number Gp-49/2001/1 in the Izn inclined drift

The maximum value of the extent of the fracture zone occurred after the third measurement. Also the number of fractures only slightly increased from 4 to 7 during the period between 10 and 20 months. The most significant increase was observed at the complete separation of the analysed roof layers, which occurred in the case when the spacing of 6 mm after the first measurement increased to 18 mm after the last measurement.

On the basis of a comparison of the measurement results obtained from the telltales and the endoscope, a conclusion can be drawn that the maximum separation obtained by the two above methods is very similar, as it is 20 mm and 18 mm respectively. However, during endoscopic measurements, no separation was indicated in the block of roof rock above the level of 2.5 m. Nevertheless, it may be assumed that in the Izn inclined drift both methods provided quite similar results.

3.2 *Comparison of measurement results obtained from telltales, endoscope and geodesic stations*

The results presented were obtained in the Cz-5 gateway located at the depth of approximately 930 m. The heading was driven in a coal seam with the thickness of 1.3–1.5 m. In the heading's roof there were mostly clay slates and arenaceous shales with slight interlayers of sandstone. The measurements were carried out in the period of about 22 months in an approx. 100-metre-long section of the heading. Support at the site consisted of ŁP9/V29 sets installed at the spacings of 1.0 m. In the examined section of the Cz-5 gateway, at the distance of 30 m, two stations for measuring low and high separations were installed, two core boreholes for endoscopic measurements were drilled and 10 stations for convergence measuring were installed.

On the basis of an analysis of changes of separation in the Cz-5 gateway with time, presented in figure 5, it may be assumed that low and high separation increased systematically during the period of about 6-7 months. After this period the changes were insignificant. The maximum high separations were -20 mm and -23 mm, whereas low separation reached the value of -18 mm.

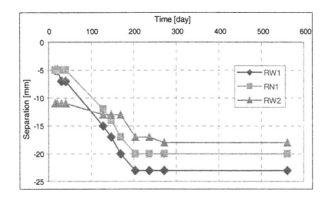

Figure 5. Change of separations in time in gateway Cz-5

The following graphs, i.e. figure 6 and figure 7, present the results of endoscopic observations for the holes located in the examined section of the analysed heading.

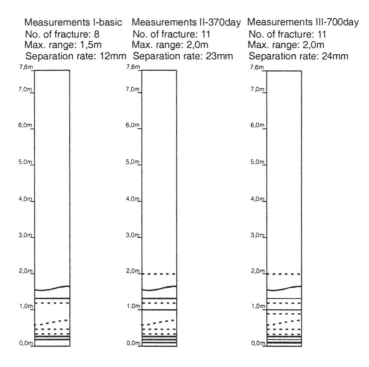

Figure 6. Results of endoscopic observations in consecutive measurement series for the hole number Gp-18/2001/1

On the basis of the measurement results obtained in two holes, i.e. Gp-18/2001/1 and Gp-18/2001/2, it may be assumed that the extent of fractures is 2.0 m and 2.5 m respectively. Generally speaking, already after the first measurement the number of discontinuities recorded on the holes' walls was bigger for the first hole. This phenomenon may be partly related to different lithology of the two holes. In the former case, clay slate occurred in the immediate roof, whereas in the latter case – arenaceous shale. The total separation at the level of 17 mm or 24 mm, which occurred in the period of almost 2 years, may be assumed as relatively insignificant. Quite similar values characterising the fracture zone between the second and third measurement, especially for hole number Gp-18/2001/1, despite the 10-month period of observation, are worth paying attention to. The measurements of convergence proved that this phenomenon results from stabilisation and suppression of rock mass movement around a heading. The ultimate number of fractures for the first hole was 11, for the second – 9.

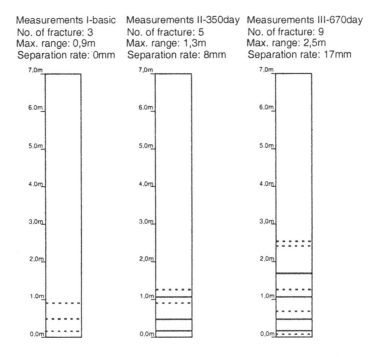

Figure 7. Results of endoscopic observations in consecutive measurements series for the hole numer Gp-18/2001/2

Figure 8 presents the subsidence of roof along the examined section of the gateway Cz-5 obtained from the measurement of convergence. The results shown here are average values obtained from 10 geodesic stations. It may be assumed from the graph below that the changes are irregular with the tendency of roof layers to subsidence, which achieved the maximum value of -32 mm. After several months the subsidence of roof stabilised to a certain level.

After the 130th day of measurements only insignificant average increase of height (9 mm) was indicated, which was related to the positive convergence (approximately 200 m) in one of the geodesic stations. In general, it may be assumed that the values obtained from the convergence measurements were higher than the ones obtained from the measurements using telltales or an endoscope. What is important in this case is the fact that the displacements indicated in the convergence measurements take into account subsidence resulting not only from the separation of roof layers, but also from elastic displacements, which do not cause separation.

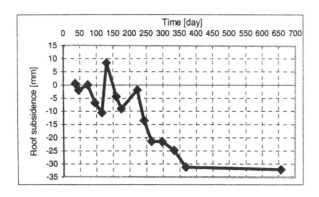

Figure 8. Average subsidence of roof along the length of the examined section

3.3. Comparison of measurement results obtained from endoscope and extensometric probes

Measurements were carried out in the B-1 inclined drift and in the 404/1 seam, along the distance of 86 m – between 500th and 586th running metre. The heading in the above-mentioned section was supported by type ŁP-V29/9 sets installed at 1 m spacings. The seam 404/1 in the area of the B-1 inclined drift had the thickness of 2.05–2.15 m and a dip of approximately 5° in the northeast direction. There was an interlayer of shale with coal lamina 0.35–0.40 m thick. The immediate roof of the heading comprised clay slate with local interlayers of arenaceous shale and sandstone, reaching up to 4.5 m. In the floor of the heading, up to 25 m, clay slate, with local interlayers of arenaceous shale was found.

Two measurement stations were installed in the control section of the B-1 inclined drift at a distance of 60 m from each other. The first station consisted of an extensometric probe and two core boreholes for endoscopic measurements, one of them drilled vertically and the other at an angle of 45° in relation to the vertical axis. In the second station there were only two core boreholes. The results of endoscopic measurements from the first and second station are presented in table 1 and table 2 respectively.

Table 1. Results of endoscopic observatons at the measurement station No. 1 in the B-1 inclined drift

Metres	Roof Hole		Inclined Roof Hole		Roof Hole		Inclined Roof Hole	
	No. of Fracture	Separation [mm]	No. of Fracture	Separation [mm]	No. of Fracture	Separation [mm]	No. of Fracture	Separation [mm]
	Basic Measurement				Control Measurement After 4 Months			
0–1	11	34	3	0	1	1	5	23
1–2	4	4	1	0	4	3	0	0
2–3	2	0	1	0	4	0	0	0
3–4	1	0	0	0	2	1	0	0
4–5	2	0	0	0	2	0	0	0
5–6	0	0	0	0	1	2	0	0
6–7	0	0	0	0	0	0	0	0
Depth	7.1 m		7.0 m		7.1 m		6.9 m	
Total	20	38	5	0	14	7	5	23

The results of observations indicate that roof layers become more and more compact with time. The number of fractures and their total separation decreases, although the extent of separated rocks increases. In the case of the measurement station No. 1, after 4 months the number of fractures decreased from 20 to 14, and the separation in the analysed 7.1 metre-long section of the roof decreased from 38 mm to 7 mm. In the case of the measurement station No. 2, after 4 months the number of fractures increased from 5 to 8, nevertheless the separation in the analysed 7.2 metre-long section of the roof decreased from 21 mm to 7 mm. The inclined roof borehole in the measurement station No. 2 behaved in a similar way: after 4 months the number of fractures decreased from 3 to 2, and separation in the analysed 7.1 metre-long section decreased from 10 mm to 0 mm. In the measurement station No. 1 the number of fractures remained unchanged (5 discontinuities), but separation increased from 0 mm to 23 mm.

Table 2. Results of endoscopic observations at the measurement station No. 2 in the B-1 inclined drift

Metres	Roof Hole		Inclined Roof Hole		Roof Hole		Inclined Roof Hole	
	No. of Fracture	Separation [mm]	No. of Fracture	Separation [mm]	No. of Fracture	Separation [mm]	No. of Fracture	Separation [mm]
	Basic Measurement				Control Measurement After 4 Months			
0–1	2	21	3	10	3	5	2	0
1–2	0	0	0	0	2	2	0	0
2–3	1	0	0	0	0	0	0	0
3–4	1	0	0	0	1	0	0	0
4–5	1	0	0	0	2	0	0	0
5–6	0	0	0	0	0	0	0	0
6–7	0	0	0	0	0	0	0	0
depth	7.2 m		7.1 m		7.2 m		7.1 m	
total	5	21	3	10	8	7	2	0

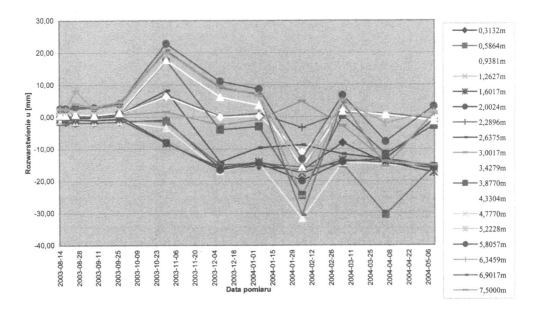

Figure 9. The results of separation measurements using extensometers in the B-1 inclined drift

Figure 9 presents the results of separation measurements carried out using an extensometric probe. The "plus" at the ratio of roof layers' displacement symbolises the subsidence of the roof (downward movement – into the heading), whereas the "minus" symbolises the compaction of rock layers.

On the basis of the graph it may be assumed that the values of displacements were insignificant, i.e. 30 mm maximum; in general they were between 0.5–15 mm. Immediately after installing the extensometer (August 14, 2003) the separation of roof rocks was very small, from 0.2 mm to 2.6 mm, but in the immediate roof compaction of rock layers occurred, which caused the decrease of separation by approximately 1 mm. It can be also noticed that there were two layers of the separating roof: the first one occurred in the immediate roof at the height of approx. 2 m., where displacements had a negative value of about -16 mm, and the second at approximately 2.8–3.23 m, where displacements also had a value of about -15 mm. The negative values occurred to the depth of 4.3 m, the roof above moved down by approximately 3 mm.

4. APPLICATION OF THE RESULTS OF MEASURING THE EXTENT OF FRACTURE ZONE

The results obtained from the measurements of separation are most often used in order to determine the amount of load exerted on the installed support or to determine the length of bolts in the case of roof bolting. Obviously, in the case of planned headings, it is impossible to measure the extent of the fracture zone, therefore prognostic methods play a crucial role in such cases. However, only numerical computations allow examining proper amount of data under (Gale, Blackwood 1987), (Jing 2003).

Computations carried out on the basis of lithology and qualities of rock layers from the Izn inclined drift, in which the measurements had been made, allowed for the determination of the extent of the fracture zone. Figure 10 presents the maps of primary stress σ_1 alongside with the determination of plasticisation zone defined on the basis of the Hoek-Brown failure criterion (Hoek, Carranza-Tores, Corkum 2002). It may be concluded from the map that the potential plasticisation zone reaches the depth of 1.2 m into the roof, i.e. the roof layer of clay slate. Much bigger extent of failure was indicated in the sidewalls and the floor, where the plasticisation zone may even reach 4–5 m.

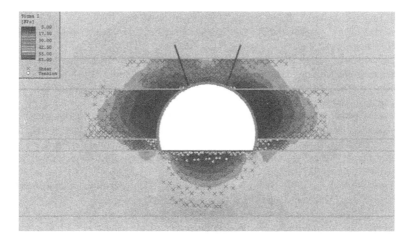

Figure 10. Map of primary stresses with plasticising zone in the Izn inclined drift
according to Hoek-Brown failure criterion

It may be claimed on the basis of the comparison between the extent of the fracture zone obtained from measurements and computations that the results are similar to each other, as in the case of the measurements, the value obtained was 0.90 m, and in the case of the computations – 1.20 m. The value estimated on the basis of computations is higher because it takes higher safety coefficient under consideration. This is advantageous, as the parameters of support selected on this basis will surely be able to carry the calculated loads.

5. CONCLUSIONS

On the basis of a coal-mine measurements of the amount of roof rock separation in a heading under development, the following conclusions may be drawn:
1. The application of several independent methods of separation measurement allows for its reliable evaluation.
2. Among the methods presented above, it is the borehole endoscope – the device indicating place, type and amount of discontinuities – that provides the most precise estimation of separation. However, to use an endoscope requires drilling a borehole and thus excludes the possibility for process automatisation.
3. Application of an extensometric probe provides numerous measurement possibilities. The measurement may be completely automated and the maximum height of 7.5 m, at which the measurement sensors can be placed, is satisfactory for the purpose of the evaluation of the heading's stability.
4. Measurements of roof displacements by means of telltales most often embrace only one block of roof rocks and thus do not indicate the changes of rock position directly above the heading.
5. As the examples quoted above indicate, the measurements of convergence can be used only as auxiliary measurements in the evaluation of roof rock separation, as they embrace the movement of the whole contour of the heading and therefore the values of displacements are higher.
6. Determination of the fracture zone in the heading's roof allows for a precise evaluation of load exerted on the installed support or length of bolts in the case of applying roof bolting.

REFERENCES

Bieniawski Z.T.: Engineering Rock Mass Classification. Wiley, New York 1989.
Chudek M.: Geomechanika z podstawami ochrony środowiska górniczego i powierzchni terenu. Wydawnictwo Politechniki Śląskiej, Gliwice 2002.
Gale W.J., Blackwood R.L.: Stress Distribution and Rock Failure Around Coal Mine Roadways. Int. Journal of Rock Mechanics and Mining Sciences & Geomechanics. Vol. 24, No. 3, 1987, pp. 165–173.
Hoek E., Carranza-Tores C., Corkum B.: Hoek-Brown Failure Criterion. 2002 edition. Proc. North American Rock Mechanics Society Meeting in Toronto 2002.
Jing L.: A Review of Techniques, Advances and Outstanding Issues in Numerical Modelling for Rock Mechanics and Rock Engineering. Int. Journal of Rock Mechanics and Mining Sciences. Vol. 40, Issue 3, 2003. pp. 283–353.
Kłeczek Z.: Geomechanika górnicza. Śląskie Wydawnictwo Techniczne, Katowice 1994.
Korzeniowski W.: Pomiar przemieszczeń górotworu przy pomocy ekstensometrów precyzyjnych. Konf. N.-T. Budownictwo Podziemne '98. Kraków 1998, pp. 263-273.
Majcherczyk T., Małkowski P., 2002: Badanie szczelinowatości skał stropowych endoskopem otworowym. Bezpieczeństwo Pracy i Ochrona Środowiska w Górnictwie Nr 2, s. 6–12.
Majcherczyk T., Niedbalski Z.: Issledovanie smieszczenij massiva gornyh porod vokrug podgotovitel'noj vyrabotki. Sbornik nauchnych trudov Nacional'nogo Gornogo Universiteta. No. 17, t. 1. Dnepropetrovsk: 2003, s. 434–441.
Niedbalski Z.: Wpływ obudowy podporowo-kotwiowej na zachowanie się wyrobisk korytarzowych w kopalniach węgla kamiennego. Praca doktorska, Kraków 2003.

International Mining Forum 2005, Sobczyk & Kicki (eds) © 2005 Taylor & Francis Group, London, ISBN 0415 375525

Technological Features of Mining in View of Geometrical and Physical Parameters of Stress Fields' Borders

Volodymyr I. Bondarenko
National Mining University. Dnepropetrovsk, Ukraine

Roman O. Dychkovskiy
National Mining University. Dnepropetrovsk, Ukraine

Successful conducting of mining in zones of influence of stress fields (BFP), the origin of which lies in the existence of various geological anomalies, depends on all-round study of the intensely complex conditions of the massive. The value of mining stress is defined by many factors; therefore there is a set of scientific views on condition of the geological environment in places of structural change of a massive.

Due to the fact that geodynamic activity in boundary parts of local fields of pressure submits to dynamic formations of higher order, there is a necessity of entering some interpretations into the existing models. They concern, first of all, character of distribution of pressure on both sides of geological anomaly.

They take into account the fact of existence in a massive near BFP of significant potential energy, which turns into kinetic energy of work of lateral strata movement. Therefore there is a zone of the lowered and increased mining pressure, respectively in downthrow and in upthrow sides of the anomaly.

A study of intensely deformed conditions of a massive was carried out with use of mine experimental researches. It's aim was to establish dependence between deformation of lateral strata and stress. Two methods of research of the massive, based on the movement of lateral breeds and changing pressure in characteristic zones are accepted for an estimation of a field of intensity.

Movement along a contour is described by Lagrange equation (1), and intensity of a field is described by Castilian equation (2) (Bezuhov 1974):

$$\delta \left[\iint_F U_0(u,v)dF - \int_S (p_x u + p_y v)dS - \iint_F (Xu + Yv)dF \right] = \delta E(u,v) = 0 \tag{1}$$

$$\delta \left[\iint_F U_0(\sigma_x, \sigma_y, \tau_{xy})dF - \int_S (p_x u + p_y v)dS = \delta E(\sigma_x, \sigma_y, \tau_{xy}) = 0 \tag{2}$$

where U_0 – specific work of internal forces; p_x, p_y, X, Y – correspondingly, planimetric and volumetric forces; δ – factor of a variation; E – total potential work of the system.

In the first case, the basis of energy change is the speed of deformation and the extent of movement. In the second - the main normal and tangential pressure show that the reason of a variation of energy E, is change of pressure σ_x, σ_y, τ_{xy}.

It is obvious that movement of lateral breeds results in stress release

$$\delta E(u,v) = \delta E(\sigma_x, \sigma_y, \tau_{xy}) = 0.$$

Therefore, carrying out of mine experimental measurements showed that there is a direct dependence between movement of lateral breeds and stress in a massive. Change of geometrical parameters of BFP on length of distribution results in adequate change of mining pressure in the zone of its influence.

Big attention was payed in the works to types of dynamic formations and their influence on intensely deformed condition at BFP. It was established that there are four types of cracks and four zones of intensely deformed conditions adjoining to the plane of an anomaly.

The scientific research encompassed only such geological anomalies, which had finished formation and were in a state of stable geodynamic activity. Preference was given to anomalies formed by tension and shear forces.

Characteristic feature of such geological anomalies is that they are in a relatively static balance, but at BFP essential potential energy is incorporated. It turns to kinetic work of layer movement due to mining operations.

Researches of the mining and geological documentation of mine fields recreate a general extend of the ramified grid of geological discontinuities. It is precisely visible, that ranged geological anomalies have strictly expressed character. Schematically fault formations can be presented through distribution of characteristic zones (fig. 1). Zone 1 is predominating area of geological discontinuities of the given site. It defines regional orientation of fields of stress and cracks in the massive of rock and ends crossing with faults of higher order. Zone 2 defines area of perpendicular tectonic anomalies and is characterized by inconstancy of amplitude and direction. Zone 3 represents formation of secondary parallel low amplitude geological discontinuities. It is formed as a consequence of unloading of pressure in units of smaller order. It has big enough distribution with faults of smaller throws and lengths (Dychkovskiy 2001).

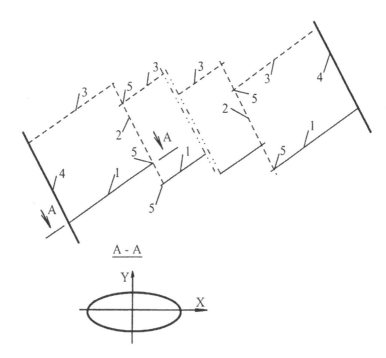

Figure 1. The circuit geological discountinuity distribution on the area of the deposit.
1 – zone of the basic geological discontinuities; 2 – zone of perpendicular tectonic geological discontinuities;
3 – zone of the secondary parallel geological discontinuities; 4 – geological discontinuities of highest ranks;
5 – points of energy accumulation

To define geometrical parameters of distribution of BFP, a single-level scanning of breeds' displacement and practical researches in conditions of mines had been undertaken. The method of geometrical similarity had established general dependence of change of low amplitude geological discontinuities on the length of distribution. The area of the discontinuity plane represents the closed space with the maximum value of amplitude in the center of the break and with gradual reduction to periphery of its distribution. The internal contour of a discontinuity plane can be approximately described by the equation of an ellipse. The horizontal axis (X) settles down in a direction of distribution of geological discontinuity, and on the ordinates axis (Y) – values of amplitude of vertical displacement of strata in a researched point are marked. Using the equation of an ellipse, it is possible to define the value of amplitude in the set point along the length of geological discontinuity from the formula

$$Y = \sqrt{1 - \frac{b^2 x^2}{a^2}} \, ,$$

where a, b – the empirical factors which are taking into account geometry parameters of geological discontinuity to actual geological conditions.

Proceeding from the structure and the forces, which were taking part in the formation of BFP, the most expedient are definitions of stress proceeding from the rank of importance. In the considered problem the vertical stress is the most important. Change of geometrical parameters of BFP along the length of distribution is adequate to change of complex conditions in a zone of their influence. According to this, there is the transformation factor (K) from amplitude of geological discontinuity to pressure. Value of pressure in upthrow (σ_n^1) and downthrow (σ_n^2) sides of discontinuity can be defined from formulae (Dychkovskiy 2001)

$$\sigma_n^1 = K_1 \sqrt{1 - \frac{b^2 x^2}{a^2}} + \gamma H \, ,$$

$$\sigma_n^2 = \gamma H - K_2 \sqrt{1 - \frac{b^2 x^2}{a^2}} \, ,$$

where K_1, K_2 – transformation factors dependent on the geological structure of breeds, sites of the researched point in space, the corner of discontinuity falling and the time factor; γH – gradient pressure in a massive (σ_0).

Condition of rock is not stationary. It depends on a place of carrying out of researches and time of a face presence at BFP. Transformation factors show the attitude of the geometrical sizes of BFP to a physical condition of a massive. Establishing the given parameters is most expedient to be done graphically as this completely displays complex specificity of a structure at BFP.

Graphic display of character of change of pressure along the length of BFP is shown on figure 2.

Using the known methodology, the load on mechanized support sets can be calculated from changes of main vertical stresses in time (Gritsko 1975).

Schematically loads at specific moments are shown on figure 3.

Knowing technical capability of support units (zone 1), it is possible to establish size and area of additional support (zone 2). If loads are greater than the technical capacity of support the given area is considered as unsuitable for mining and is not included in the layout of development works (zone 3).

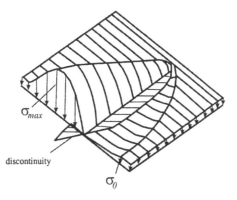

Figure 2. Character of change of stress in boundary parts.
σ_{max} – maximum vertical stress; σ – stress of an equilibrium condition

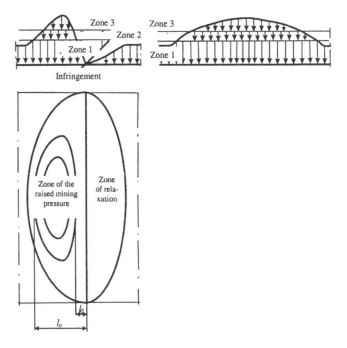

Figure 3. Definition of loads in a zone of influence of BFP.
l_0, l_n – the distances describing a zone of installation of additional support

The size of necessary bearing capability of face support depends on loading of lateral breeds and character of mining stress. It changes in time and in space. Especially it is dependent on BFP. To achieve normal operating conditions it is necessary to observe the condition (Dychkovskiy 2002):

$$R_{кp} \geq P ,$$

where $R_{кp}$ – support reaction, KN/m^2; P – support load, KN/m^2.

Bearing capability of support depends on technological capability of mechanized longwall sets and additional support:

$$R_{кр} = \Sigma R_{кр.м.к.} + \Sigma R_{доп},$$

where $R_{кр.м.к.}$ – bearing capability of the mechanized sets, KN/m^2; $R_{доп}$ – bearing capability of additional support, KN/m^2.

Definition of the site and sizes of reaction of additional support is the decision of a problem with two variables: the size of reaction of additional support ($R_{д.к.i.} = var$) and distance from the place of its installation to discontinuity ($l_i = var$). For the solution of the task we shall limit the researched area, proceeding from come loadings to the zone of additional support (fig. 4). The maximum load in a researched time interval does not exceed total technical support capabilities of the mechanized support set and additional props.

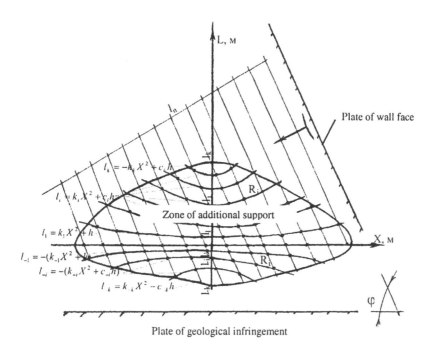

Figure 4. Circuit of management of mining stress at BFP in a zone of additional support

Let's impose on the researched area a grid of the Cartesian coordinates as follows: an horizontal axis – parallel to the trace of geological discountinuity; an axis of ordinates – perpendicularly to BFP. The point of zero (crossing of axes) corresponds to the maximum load, falling to the zone of additional support. Placement of additional support props is defined proceeding from the loadings, falling to the area of i-ing site and is characterized by three parameters: distances between props of additional support along the length of the column (l_n), distances between props of additional support along the length of longwall face (l_i) and also size of bearing capability of additional support ($R_{д.к.i.}$). Negative values on the axis of ordinates (l_{-i}) specify placement of additional support from displacement up to the maximum load, positive (l_i) from the maximum load up to the border of the zone of additional support.

67

The distance between units of additional support along the length of the column is defined, proceeding from technical capabilities of the mechanized support set, from the formula

$$l_n = rn \, tg \, \varphi,$$

where r – width of extraction by the cutting loader, m; φ – angle of intersection between the longwall and geological anomaly, degree; n – frequency of installation of additional support units on length of the column.

The value should be a multiple of the number of cycles of face cuts; it is defined by expression

$$n = \frac{P_i - R_{\text{м.к.}}}{R_{\text{д.к.i}}},$$

where P_i – load on support of i-ing site, MPa; $R_{\text{м.к.}}$ – bearing capability of sections of mechanized support, MPa; $R_{\text{д.к.i}}$ – bearing capability of additional support units, MPa.

Bearing capability of additional support depends on the type of units used, their quantity and placement on the area of i-ing a site:

$$R_{\text{д.н.i}} = n_R R_{\text{д.н.max}},$$

where $R_{\text{д.н.max}}$ – maximum bearing capability of one additional support unit, MPa; n_R – number of additional support units in the area of i-ing site.

The final number of frequency of installation of additional support units is approximated to the nearest smallest value.

For efficient control of mining stress in the longwall it is necessary to observe the following condition:

$$P_i \leq R_{\text{м.к.}} + \frac{R_{\text{д.к.i}}}{l_i s \cos \varphi},$$

where l_i – distance between units of additional support along the length of longwall face, m; s – width of longwall face, m.

Accordingly the required bearing capability of additional support, proceeding from loading of lateral breeds, is defined by expression

$$P_i \leq \frac{R_{\text{м.к.}} + R_{\text{д.к.i}}}{l_i s \cos \varphi}.$$

The distance between units of additional support along the length of longwall face is calculated from expression

$$l_i = \frac{R_{\text{м.к.}} + R_{\text{д.к.i}}}{P_i s \cos \varphi}.$$

On the other hand the value l_i can be mathematically determined. Change of distance of additional support units in the area of a zone of additional support is most precisely reflected on the basis of a curve of the second order with parabolic dependence. As placing of additional units is planned in spaces between sections of mechanized support the free member of the equation should be multiple to the step of support installation (h). The minimum distances ($l_1 = h$) and ($l_{-1} = h$) should correspond to the maximum load on the horizontal axis. The zone where observance of this condition is impossible, discontinuation of extraction works (bypass zone) is considered. In such a case full or partial dismantling of the mechanised support and subsequent re-installation of the missing sections behind the bypass zone is done.

The distance between the units of additional support is defined from a maximum load aside a massive by a curve of the second order from expressions

$$l_1 = k_1 X^2 + h,$$

$$l_i = k_i X^2 + c_i h,$$

where h – a step of installation of mechanized support sections, m; c – the factor which is taking into account a step of installation of additional support units.

Accordingly, the placement of units from the maximum load up to the plane of discontinuity is defined by expressions

$$l_1 = -(k_{-1}X^2 + h),$$

$$l_i = -(k_{-i}X^2 + c_{-i}h).$$

The zones of additional support is defined proceeding from crossing of two hyperboles:

$$l_k = -k_k X^2 + c_k h,$$

$$l_{-k} = k_{-k}X^2 - c_{-k}h.$$

The actual extreme placement of additional support units in a zone of additional support is defined by crossing each line of installed support with the contour. Mathematical display is represented as the solution of complex systems of the equations

$$\begin{cases} l_1 = k_1 X^2 + h \\ \downarrow \cdots\cdots\cdots \downarrow \\ l_{k-1} = k_{k-1}X^2 + c_{k-1}h \\ l_k = -k_k X^2 + c_k h \end{cases},$$

$$\begin{cases} l_{-1} = -k_{-1}X^2 - h \\ \downarrow \cdots\cdots\cdots\cdots \downarrow \\ l_{-k+1} = -k_{-k+1}X^2 - c_{-k+1}h \\ l_{-k} = k_{-k}X^2 - c_{-k}h \end{cases}$$

The best manner of additional support is installing wooden or hydraulic props in spaces between sections of the mechanized sets. At achievement of critical load when the value of mining stress exceeds technological support capacity, the site is considered unsuitable for mining and full or partial re-installation of the mechanised support is done.

On the basis of the research it may be said that there is an opportunity of acceptance of technological solutions in conducting mining works adequately in areas of changes of geometrical and physical parameters of a massive at BFP. Procedures of situational management in zones of structural changes mean overcoming borders of stress fields at various spatial positions of geological discontinuities. Practical approbation of introducing the results in mine conditions prove this approach to management of massive at BFP.

REFERENCES

Bezuhov O.V., Luzin O.V.: Methods of Elasticity and Plasticity Theory to the Decision of Engineering Problems. Moscow, Higher School 1974, p. 115–128.

Dychkovskiy R.O.: Mathematical Modeling of Geometrical Parameters Influence of Intense Fields on Intense-Deformed Condition of Massif. Proceedings of the International Symposium on GeotechnologicalI Issues of Underground Space Use for Environmentally Protected World. Dnepropetrovsk, Ukraine, 26–29 June 2001, p. 167–170.

Dychkovskiy R.O.: Analytical Definition of Pressure in the Massive of Flat Layers on a Joint of Changing Structures. Geomethanics, Dnepropetrovsk: IGTM NAC of Ukraine 2001. No. 29, p. 112–118.

Gritsko G.I., Vlasenko B.V.: The Common Principles of an Estimation of an Intense Condition of the Massive by an Experimental – Analytical Method of Rocks Mechanics. Analytical Methods and Computer Facilities in Rocks Mechanics. Novosimbirsk, 1975, p. 109–116.

Dychkovskiy R.O., Gajdaj A.A.,: Definition of Loading on Wall Support at Pressure Fields Borders on Flat Layers. Scientific Bulletin of HMU 2002, No. 1, p. 30–33.

International Mining Forum 2005, Sobczyk & Kicki (eds) © 2005 Taylor & Francis Group, London, ISBN 0415 375525

Optimization of Interaction in the "Anchor – Rock" System

Irina A. Kovalevska
National Mining University. Dnepropetrovsk, Ukraine

Viacheslav V. Porotnikov
State company "Alexandriacoal". Ukraine

The Ministry of Fuel and Energy of Ukraine presented a task to deacrease the share of expenses for construction and operation of underground mining workings to 6–8% of the total cost of coal. According to this task a special program of support by roof bolt systems is developed and actively used and its performance is realized through a system of the state programs of step-by-step and comprehensive application of bolt support. Constructions and methods of rational parameters of tubular anchors (Bondarenko 2004) for effective consolidation of weak strata are designed and developed at the National Mining University. Today, the problem of intense – deformed condition of "tubular anchor TA2 – rockwall of the hole" system is solved, the components of stress and their basic interrelations with the most influencing geometrical, mechanical and other parameters of system elements are determined.

Researches on transformation (optimization) of the stress field so that to provide the maximum reaction of resistance Q of tubular anchor TA2 in a pliable operating mode at consolidating weak rocks are executed on this basis. It is the main criterion when estimating the overall performance of tubular anchor TA2. Researches established that the maximum reaction of resistance Q of an anchor in a pliable mode is achieved at required restriction of narrowing of its cross-section due to the action of filler. In result on the external (loaded) end of an anchor (at its active length l_a) the maximum tension in an anchor reaches the best value equal $\beta_1\sigma_T$. Therefore the main criterion of the maximum reaction of anchor TA2 on an axis Z we shall write down as

$$\text{at} \quad Z = -\frac{l_a}{2} \qquad \sigma_z = \beta_1\sigma_T \tag{1}$$

where β_1 – factor of hardening of material in a metal pipe at an explosive way of its fastening in hole; σ_T – limit of plasticity of steel.

For realization of condition (1) it is necessary to formulate additional criteria.

First, at the limited active length of anchor l_a in conditions of weak strata (small stress P_{ul} upon contact "tubular anchor TA2 – hole rockwalls") a situation can arise when the reaction of anchor resistance Q cannot achieve durability Q_{max} on break of the metal pipe. This case takes place, when:

$$l_a < \frac{d_{ш}^2 - d_1^2}{4d_{ш}} \cdot \frac{\beta_1\sigma_T}{f_{тр}R^\infty} \tag{2}$$

where $d_{ш}$ – diameter of hole; d_1 – internal diameter of an anchor; f_{mp} – factor of friction of metal on rock; R^{∞} – settlement long resistance of rock on compression in a massive.

In practice the length of an anchor, as a rule, does not exceed 2,5 m, and its active length is in limits up to 2–2,4 m. If active length of an anchor l_a calculated by the formula (2) exceeds the specified value with a view of achievement of the greatest possible durability of fastening of an anchor in the given conditions the additional criterion is formulated so: along all length of the anchor narrowing of its cross-section should be completely excluded, that is

$$at \quad \sigma_z = \frac{4d_{ш}}{d_{ш}^2 - d_1^2} \cdot l_a f_{тp} R^{\infty} \qquad P_{ш} = R^{\infty} \tag{3}$$

Realization of this criterion is possible in two ways: application of a filler which does not expand when setting; application of a filler with volumetric expansion at hardening a sand-cement mix that allows to create initial pressure P_3^H upon contact "filler – metal pipe".

In case of application of filler not extending in volume formula for calculation of its minimally necessary module of deformation E_3 is received:

$$E_3 \geq (1-\mu_3) \left[4,24 \left(\frac{d_1}{d_{ш}} \right)^{-4,4} - 3,99 \right] \cdot 10^4, \quad MPa \tag{4}$$

where μ_3 – Paulson factor of filler.

The formula (4) is received on the basis of the numerical analysis of an intense condition of the system "filler – metal pipe – rockwalls of hole" with application of methods of the depressive analysis.

In case the required value E_3 obtained by calculations is inconvenient for receiving in existing usually used sand-cement stones, volumetric expansion at setting is necessary to be applied. The sand-cement mixes creating initial pressure P_3^H on contact "filler – metal pipe". Researches of intense – deformed conditions of system "filler – metal pipe – rockwalls of a hole" and processing of the received results by methods of the depressive analysis have allowed to receive formula for calculating minimum necessary initial pressure P_3^H excluding cross-section narrowing of a metal pipe:

$$P_3^H \geq 0,215 \beta_1 \sigma_T \left(\frac{d_{ш}^2}{d_1^2} - 1 \right) - \left[100 + 124 \left(\frac{d_1}{d_{ш}} \right)^{9,3} \right] \frac{E_3}{(1-\mu_3)E}, \quad MPa \tag{5}$$

where E – module of elasticity of steel.

Secondly, at certain parities of mechanical properties and geometrical parameters of elements of "filler – metal pipe – rockwalls of the hole" system in weak rock, a situation when the size of active length of an anchor l_a allows to realize the maximum bearing capability Q_{max} of an anchor, i.e. when the inequality (2) is not carried out, is quite possible. In this case the criterion (3) can be softened in a direction of an assumption of certain narrowing of cross-section of tubular anchor TA2 under the influence of axial tension. Really, if the active length l_a of an anchor on an inequality (2) the tubular anchor can achieve the maximum bearing capability Q_{max} even at stress decrease $P_{ш}(\sigma_Z)$ on sites of

72

action of the axial pressure σ_Z which are coming nearer to the maximum value. Then it is possible to write down less rigid criterion as

$$\text{at} \quad Z = -\frac{l_a}{2} \qquad\qquad P_\text{ш} = \left(P_\text{ш}\right)_{\min} \tag{6}$$

$$\text{at condition} \quad \frac{4d_\text{ш}}{d_\text{ш}^2 - d_1^2} \cdot f_\text{тр} \int_{-\frac{l_a}{2}}^{\frac{l_a}{2}} P_\text{ш}(Z)\,dZ = \beta_1 \sigma_\text{т},$$

where $(P_\text{ш})_{\min}$ – minimally allowable value of stress on "metal pipe – rockwalls of the hole" contact on its external end.

Researches established that along the length of an anchor stress $P_\text{ш}(Z)$, as well as tangents stress $\tau_{rz}(Z)$, is distributed according to exponential dependence:

$$P_\text{ш}(Z) = R^\infty \exp(-n_1 Z) \tag{7}$$

where n_1 – dimensionless factor determined by a parity of parameters of elements of "filler – metal pipe – rockwalls of the hole" system.

Then according to criterion (6) at the account of the equation (7) the transcendental equation for definition of the required value of factor n_1 at which tubular anchor TA2 achieves the maximal bearing capability is received:

$$n_1 = -\frac{d_\text{ш}^2 - d_1^2}{4d_\text{ш}l_a f_\text{тр}} \ln\left(1 - n_1 \frac{\beta_1 \sigma_\text{т}}{R^\infty}\right) \tag{8}$$

The numerical solution of the equation (8) has allowed to establish the following correlation dependence:

$$n_1 = -1{,}89\left(\frac{d_\text{ш}^2 - d_1^2}{d_\text{ш}l_a f_\text{тр}}\right)^{1,5} + \left(0{,}99 + 14{,}3\frac{d_\text{ш}^2 - d_1^2}{d_\text{ш}l_a f_\text{тр}}\right)\frac{R^\infty}{\beta_1 \sigma_\text{т}} \tag{9}$$

Rational value of factor n_1 is determined by a complex of parameters of "filler – metal pipe – rockwalls of the hole" system. Studying the interrelation of these parameters with the help of the approved methods of the depressive analysis has allowed receiving expression for calculation of the minimal necessary value of the module of deformation E_3 of the filler providing the maximal bearing capability of tubular anchor TA2 without use of effect of volumetric increase of filler at setting:

$$E_3 \geq \left(1 - \mu_3\right)10^4\left[135\left(1 - \frac{d_1}{d_\text{ш}}\right)^{1,8} - 14{,}7\left(\frac{E}{E_\text{п}}\right)^{0,8}\left(\frac{d_\text{ш}}{d_1}\right)^{6,5} n_1\right], \quad \text{MPa} \tag{10}$$

where $E_\text{п}$ – the module of elasticity of rock.

If we achieved the module of deformation E_3 of a filler required by calculation in usual sand-cement stones that is not obviously possible, as a filler it is necessary to apply extending at strengthening the sand-cement mortals creating initial pressure P_3^H on "filler – metal pipe" contact not less than value:

$$P_3^H \geq \left(1 - \frac{d_1^2}{d_{\text{ш}}^2}\right)\beta_1\sigma_{\text{т}}\left(0,3 - \frac{4,2}{1-\mu_3}\left(\frac{d_1}{d_{\text{ш}}}\right)^{5,6}\frac{E_3}{E} - \left\{10,5\left(\frac{d_1}{d_{\text{ш}}}\right)^{3,2} + \left[3,1 - 2,55\left(\frac{d_1}{d_{\text{ш}}}\right)^{1,2}\right](1+\mu_n)\frac{E}{E_n}\right\}n_1\right)$$

(11)

where μ_Π – Paulson factor of rock.

The numerical analysis of size of minimum necessary initial pressure P_3^H on the basis of which dependences (5) and (11) are constructed, has shown that its value does not exceed, as a rule, 40–50 MPa also is quite commensurable with strengthening characteristics of a sand-cement stone on a basis of widely used marks of Portland cement.

As it has been specified earlier, initial pressure P_3^H on "filler – metal pipe" contact arises due to increase in volume at strengthening of the sand-cement mix with expanding additives. Relation between initial pressure P_3^H and factor of volumetric expansion K_3 of filler is determined from a condition of flat deformation of cross-section section of filler according to classical parities (Pisarenko 1979):

$$K_3 = 2 \cdot 10^2 \left(1 - \mu_3 - 2\mu_3^2\right)\frac{P_3^H}{E_3}, \quad \%$$

(12)

where K_3 – factor of volumetric expansion of filler, expressed in percentage.

The dosage of expanding additives in the sand-cement mortal of filler is made according to the settlement value K_3 determined under formulae (5), (11) and (12).

By way of optimization of the intense – deformed conditions of "filler – metal pipe – rockwalls of the hole" system it is necessary to consider the problem of limiting condition of materials of filler and pipe. As to the limiting condition of a metal pipe this question is in details investigated in work (Simanovich 1982) where it has been established, that process of performance of a condition of durability of anchor TA1 without filler at a stage of plasticity of steel is automatically adjusting and is carried out automatically: axial stretching of component σ_Z tubular anchor TA1 increases equally for size of reduction tangential compressing components σ_θ and pressure $P_{\text{ш}}(\sigma_Z)$ on "rockwalls of the hole" contact becomes equal to zero when in this section of a component σ_Z achieves the greatest value $\beta_1\sigma_{\text{т}}$.

The material of a filler also is in three-dimensional intense condition: radial σ_r^3 and tangential σ_θ^3 components are compressing, axial a component σ_z^3 – stretching. Because of low resistibility of a sand-cement stone to stretching efforts in filler on axial coordinate Z there is a lot of cross-section sections of break virginity, where $\sigma_z^3 = 0$. According to the Colon-Moro theory of durability and parities (Simanovich 1982) it is possible to write down, that minimum component is equal $\sigma_{\min} = \sigma_z^3 = 0$, and maximum compression component is equal $\sigma_{\max} = \sigma_r^3 = \sigma_\theta^3 = P_3^H$. Then the condition of durability of filler has a simple relation:

$$P_3^{\text{н}} \leq \sigma_3^{\text{сж}} \tag{13}$$

where $\sigma_3^{\text{сж}}$ – strength of a material of filler on one axis compression.

Thus, as a result of optimization of the intense – deformed conditions of "tubular anchor TA2 – rockwalls of the hole" system expressions for calculation of rational mechanical characteristics of the filler providing the maximum reaction of tubular anchor TA2 in a pliable operating mode at hardening of weak rocks are received.

REFERENCES

Bondarenko V.I., Simanovich G.A., Kovalevska I.A., Porotnikov V.V.: The Effectiveness of Support of Weak Ruck with TA2 Tubular Anchors. International Mining Forum, New Technologies in Underground Mining, Safety in Mines 2004, p. 43–46.
Pisarenko G.S., Agaev V.A., Kvitka A.L. et al.: Resisting of Materials. Higher School, Kyiv 1979, p. 643.
Simanovich G.A.: Interaction of Breed Massif with Pipe Bolts and Development of Calculation Method of Its Parameters. Dnepropetrovsk, IGTM AC USSR 1982, p. 193.

International Mining Forum 2005, Sobczyk & Kicki (eds) © 2005 Taylor & Francis Group, London, ISBN 0415 375525

Determining the Design and Parameters of Straight Cylinder Burn Cut in Development Workings

Garry G. Litvinsky
Donbass State Technical University (DonSTU). Alchevsk, Ukraine

Pavel Shulgin
Donbass State Technical University (DonSTU). Alchevsk, Ukraine

ABSTRACT: The basic information about cylinder cuts used in mine works are given. The advantages of cylinder cuts are shown. The purpose of research – development of design and determination of parameters of straight cylindrical burn cut. Design features of cylinder cut and a principle of its work at explosive destruction of rocks are considered. Cut parameters are investigated, calculation formula for its radius depending on major factors is offered. Recommendations for practical application of cylinder cut are given, its advantages are shown.

KEYWORDS: Drilling and blasting operations, cylinder burn cuts, design of cylinder cut, parameters of bore-holes, calculation methods, factors' valuation, application recommendations

1. IMPORTANCE OF THE PROBLEM FOR BLASTING OPERATIONS AND PURPOSE OF THE INVESTIGATION

In contemporary underground mines development is, as a rule, done with the help of drilling and blasting operations. Last years more than 80% of total amount of the main development tunnels were developed in this way. That's why increasing the efficiency of drilling and blasting operations has the opportunity to greatly raise the rapidness and to decrease the cost of excavation construction.

The most important element in the process of drilling and blasting operations is burn cut, which in many respects predetermines the face advance and quality of crushing of rock. From numerous kinds of cuts available at present, the most progressive are straight cylinder burn cuts, made by bore-holes bored perpendicular to the surface of hard-rock heading (slot cut, prismatic, spiral-stepped cuts, dual stepped cut, cylinder cut with cleaning-up bore-hole and others) (Smirnyakov 1989; Savelev 1982; Mindeli et al. 1978; Langefors, Kilstrem 1968; Pokrovskiy, Fedorov 1957; Mindeli 1966; Guyboroda, Nalisko 1994; Vlasenko, Shevcoh, Gudz, Shyupkin 1998; Shpunt 1992).

Out of all known structures cylinder cuts are characterized by high capacity for work, universality of application, stable parameters, and simplicity of orientation in space (Guyboroda, Nalisko 1994; Vlasenko, Shevcoh, Gudz, Shyupkin 1998; Shpunt 1992). However, the existing designs of cylinder cuts have a number of disadvantages: complicated patterns in hard-rock headings, necessity of using bore-holes of various length, in the number of cases they require electric detonators with a big range of delays to be available, stimulate increased rock ejection from the face and high probability of support damage. Almost all cuts have the important disadvantage – absence of methods allowing determining their parameters and conducting necessary calculations.

The purpose of this research is development of new design and method of calculative determination the parameters of cylinder cut, which eliminate all the limitations listed above.

2. DEVELOPMENT OF THE DESIGN OF CYLINDER CUT

On the basis of analysis of existing designs of cylinder cuts it is possible to conclude, that in general they do not satisfy the main purpose of using a cut – the optimal concentration of blast energy and creating the cut cavity of maximum volume and extension.

As optimal blast energy concentration in rock we understand such minimum quantity of explosive material and its particular space location, which according to strength characteristics of rock provide its crashing in the required volume and rock ejected without destroying the installed support.

For this purpose it is necessary to place around the central borehole as many cylinder cuts so that the distances between cuts are equal. Only in this case blast waves from all charges can reach the cut centre and each neighbouring borehole at the same time to provide the required high level of rock destruction.

Based upon the geometrical construction of holes' location and movement of blast wave it is possible to be convinced that the maximum volume of the destroyed rock with minimal charge of explosives could be obtained by using cylinder cut designed in the form of a central bore-hole, around which other bore-holes are placed at equal distances between them and the central bore-hole. As follows from the geometry of such positioning, these boreholes have to be located on a circle with the radius equal to distance between neighbouring boreholes, and their quantity is equal to six. Thus, the cut contains only seven boreholes, six of which are placed around the central borehole and form a cylindrical surface.

Such positioning of boreholes (fig. 1) creates extremely high and equal concentration of explosive energy. Another quantity of boreholes results in irregularity of explosion energy concentration inside the cylindrical cut. It is especially important for the case when it is necessary to break hard rock as cylindrical cut allows achieving the necessary crushing of rock and index of borehole use, simultaneously observing the requirements of safety rules with regard to restriction of minimum allowed distance between bore-holes.

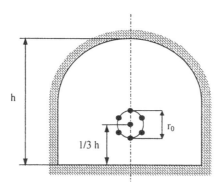

Figure 1. Positioning of cut at the face

Operation of cylindrical cut consists in the follows (fig. 2). Charges 5 are blasted first (e.g., electric detonators 0-delay) in cut boreholes 2. At their simultaneous blast the cylindrical explosion waves due to being placed around boreholes 2, simultaneously meet at the central borehole 1, interact and are reflected from its walls, which results in destruction of rock within the inside perimeter of cut 6 and its bottom part 7. Outside the circle of boreholes 2 the zone of radial and tangential fractures 8 is created. The gaseous products of explosion from charges 5 throw out the destroyed rock, thus resulting in creating a cut hollow 6 with a depth of 0,7–0,8 of the drill length L_3. The bottom rock 7 of the cut hollow 6 is compressed around the central forward borehole, forming a stemming of increased density, and the charge 3 is outside the zone of influence of the compressive action of charges 5.

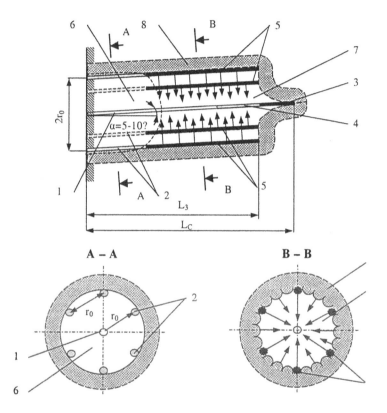

Figure 2. Principle of operation of cylinder cut

Next, due to blasting of charge 3 in the central bore-hole 1 (e.g., by electric detonator with 15 ms delay), the gaseous blast products not able to escape into the air due to compression of the rock complete mechanical work on breaking and ejecting the rock left in the bottom part 7 of the cut hollow 6. In that way they finally form the cut hollow of the depth equal to drill length L_3, which provides the effective work of all other bore-holes in the face with high index of bore-hole use.

3. DETERMINING PARAMETERS OF THE PROPOSED CUT

Let's limit ourselves to determining expedient parameters of proposed cylinder cut. The most important of them because of the simplicity of the cut's geometry is the radius of the circle on which the boreholes are located. In this case the formula generally accepted in calculation process of parameters of blasting works is used (Langefors, Kilstrem 1968; Pokrovskiy, Fedorov 1957; Mindeli 1966), the direct relation between the volume of broken rock and consumption of explosives:

$$Q_B = q \cdot V \tag{1}$$

where Q_B – weight of explosives for the whole cut, kg; V – cut volume, m^3; equal to:

$$V = \pi \cdot r_0^2 \cdot L_3 \tag{2}$$

r_0 – radius of the cut's hollow, m; L_3 – drill length for the cycle, m; q – specific consumption of explosives (determined by formula of Prof. N.M. Pokrovskiy (Mindeli et al. 1978), kg/m^3.

Weight of explosives is calculated from the weight of charges in six cut boreholes:

$$Q_B = 6 \cdot Q_Z \tag{3}$$

where Q_Z – charge weight in a single cut bore-hole, equal to:

$$Q_Z = \Delta \cdot S_3 \cdot (L_3 - l_Z) \tag{4}$$

$l_Z \geq 0,5$ m – minimum length of stemming; Δ – density of explosives, kg/m^3; S_3 – sectional area of charge, m^2; L_3 – drill length for cycle, m.

Substitute into expression (1) equations (2), (3) and (4), solve it relatively to r_0, and then get formula for determining the radius of cylinder cut:

$$r_0 = \sqrt{\frac{6 \cdot Q_Z}{\pi \cdot L_3 \cdot q}} \tag{5}$$

As it is seen from the formula cut radius depends on the drill depth, specific consumption of explosive material (and hence from the rock's strength properties, capacity of explosives for work and transverse section of mine working).

Let's for example determine the values of each of the 5 factors included in the formula (fig. 3). For this purpose we shall take values of elasticity indexes of each factor. Elasticity index is defined as a degree of change of relative variable value of the initial factor.

4. INVESTIGATING THE INFLUENCE LEVEL OF FACTORS AND THEIR VALUATION

From the diagrams (fig. 3) it is seen how significant are changes of the radius of cylinder cut depending on the explosives' strength (from 0.3 to 1.1 m) and on the rock's strength (from 1.3 to 0.2 m). Average elasticity index of explosives' strength is 0.52, and for rock's strength (strength on Prof. M.M. Protodjakonov's scale) this index is equal to -0.58. Minus shows that increasing the factor causes the variable value (radius of cylinder cut) to decrease. Consequently the elasticity index for excavation area is 0.27, and for drill length it is -0.43. Thus, the most significant factors are the rock strength and explosives' strength.

Comparison of the degree to which various factors influence the radius of cylindrical cut is given in table 1.

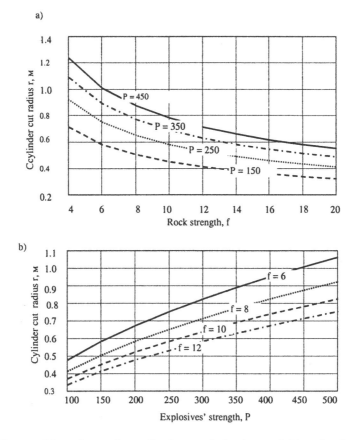

Figure 3. Diagram of dependence of cut radius from explosives' strength (a), and rock strength (b)

Factor valuation has shown that to increase the efficiency of cut in hard rock it is probably necessary to use more efficient explosives, and if the index reaches the limit to change to shorter drill length L_3.

5. RECOMMENDATIONS FOR PRACTICAL APPLICATION

When using cylinder cut one should choose its parameters guided by the following recommendations (fig. 2):

1. The central borehole 1 must be drilled deeper than drill depth L_3, according to the empiric formula:

$$L_C = L_3(1+0,1\cdot\sqrt[3]{f})$$ (6)

2. In the central frontal borehole 1 it is necessary to place stemming 4 of minimum length allowed by safety regulations, and charge 3, determined by the formula:

$$Q_C = Q_0 \cdot L_3 \cdot 0,1\cdot\sqrt[3]{f}$$ (7)

3. To decrease the distance of rock ejection and decreasing the probability of support damage place the centre of cut on 1/3 height of the face from the floor (fig. 1), and all bore-holes should be drilled by slightly up with an angle of 5–10 degrees to the floor level. Placing the cut lower is not recommended because in this case it will be placed in the zone where almost all strain deformations in front of the face are absent. It considerably decreases the efficiency of a cut. Placing the cut higher than one-third height of face above the floor results in enormous ejection of rock from the face and increases the probability of support damage whereas efficiency of creating the cut hollow is not increased. Placing the boreholes at angles greater than 5–10 degrees to the tunnels floor is not recommended because it complicates the pattern and the drilling process. Boreholes at smaller angle are ineffective.

Table 1. Elasticity index and values of main factors (5)

Influencing factor	Elasticity index K_e	Factor's value $R = |K_e/K_{сред}|$
Rock strength, f	-0,58	1,3
Explosives' strength, P	0,53	1,17
Drill length, L_3	-0,43	0,94
Area of the working, S	0,27	0,6

CONCLUSIONS

The proposed construction and calculation methods for cylinder cut allow to consider geological and technical conditions when choosing all cut parameters; provide high efficiency of cut hollow and blasting with high index of bore-hole efficiency; decrease the probability of causing damage to the support by the blast and limit rock ejection from the face.

The recommended application area of cylinder cut – horizontal and inclined mine workings. The proposed construction of cut would be especially effective for driving of workings in hard rocks (index of hardness on Protodjakon's scale of more than 6–7). Using the proposed cut will allow to increase technical and economic characteristics of blasting operation in mine works.

REFERENCES

Smirnyakov V.V. et al.: Technology of Construction of Mines. M. Nedra 1989, p. 289–292.
Savelev U.Y. et al.: Intrusion of a Burn Cut with Cylinder Cut with Cleaning-up Bore – Charge at Realization of Horizontals Workings. Mining Magazine 1982, №3, p. 53–54.
Mindeli E.O. et al.: Complex Research of Action of Explosion in Rocks / (E.O. Mindeli, N.F. Kusov, A.A. Korneev, G.I. Marcinkevich). M. Nedra 1978.
Langefors U., Kilstrem B.: Modern Engineering of a Blasting of Rocks. M. Nedra 1968.
Pokrovskiy G.I., Fedorov G.S.: Action of Impact and Explosion in Deformable Environments. M. Nedra 1957, p. 183.
Mindeli E.O. Drilling-and-Blastings at a Underground Mining of Mineral Resources. M. Nedra, 1966, p. 555.
Guyboroda V.N., Nalisko N.N.: Direct Stepping Kerf. Coal of Ukraine 1994, № 7, p. 25–26.
Vlasenko V.Y., Shevcoh N.R., Gudz A.G., Shyupkin N.N.: Walking Cuts. Coal of Ukraine 1998, № 2, p. 20–22.
Shpunt V.I.: Cylindrical Stepping Kerfs. Coal of Ukraine 1992, № 1, p. 24–29.

International Mining Forum 2005, Sobczyk & Kicki (eds) © 2005 Taylor & Francis Group, London, ISBN 0415 375525

Potencial for Use of Pneumatic Constructions in Underground Mining

Volodymyr I. Buzilo
National Mining University. Dnepropetrovsk, Ukraine

Volodymyr S. Rahutin
National Mining University. Dnepropetrovsk, Ukraine

Volodymyr P. Serdjuk
National Mining University. Dnepropetrovsk, Ukraine

The share of manual labour in underground mining remains very substantial. The potential of the use of hydraulics in traditional sense is already exhausted to a large extent and will not provide further help in decreasing the volumes of manual labour.

The collaborators of the National Mining University (Dnepropetrovsk, Ukraine) offered a new principle of mechanization for processes of underground mining. They first developed a theory which served as a base for the practical application of pneumatic soft shell constructions in underground mining.

Soft shells may be described as impermeable bags made from durable technical fabrics.

The principle of work of soft shells consists in expanding them in cavities thhrough the application of compressed air. Thus the lateral surfaces of shells create initial support to the rock surface carrying out the work of containing rocks.

Initial hold is determined by the work over the area of contact of a soft shell with containing rocks on surplus pressure in a cavity. So, for example, areas of contact 1 m^2 and surplus pressure 0,4 MPa, will result in initial hold of 400 kN.

Soft shells have a number of advantages in comparison with other technical means for mechanization of processes of underground mining works.

- Small thickness in initial position (3–10 cm), that allows to place soft shells in small spaces between support and surface of rock.
- Multiple leafing (for example, from 0,3 cm to 600 cm), that allows a wide range of their use.
- Large area of contact with containing rocks and small specific pressure (0,05–0,5MPa) that allows to use soft shells for support of weak rocks (clays), not damaging them.
- Simplicity of construction (absence of valves, hydro blocks and other).
- Small weight resulting in minimal labour output ratio of assembling-dismantling works.
- Good filling of spaces with uneven surfaces, including domes.

Soft shells are made in two types: vulcanized and mat. A rubber-covered cord and raw rubber are the raw materials for making of vulcanized shells (material from which overlays are made).

Mat shells are made from artificial materials – rubber-covered fabrics, poliester, sleeve and kapron fabrics. The technology of their manufacturing is substantially simpler than the vulcanized shells, because it does not require press form, formations – vulcanizations, overheated steam and other.

Pneumatic soft shells are used for mechanization of processes of underground mining works in coal and ore mines, during building of underground passages and other underground constructions.

In coal mines serially made pneumatic props are used. They replaced wooden props, mainly during development of thin steep seams, providing economy of timber usage, promoting labour productivity and work safety (Pneumatic Constructions... 1983), (Nekrasovskyy 1975).

During the underground extraction of ores by means of systems of development with complete fill of the mined-out space, systematic fill dams are erected which hinder penetration of liquid fill material into the working excavations. At present the dams are made from timber, concrete, reinforced concrete or combination of these materials. After consolidation of fill the dams, although not needed any more, forever remain in the place of installation.

Dams constructed from soft shells in comparison with traditionally used have a number of substantial advantages, among them: potencial for multiple use; quickness of erection and dismantling; good adapting to the inequality of walls of excavations; universality of construction.

The industrial tests of pneumatic backfill dams showed their efficiency (Khmarskyy 1982).

During building of underground passages large volumes of support works are executed. In support space between supports the rock surface breaks. For withholding of loose rock in support space between support sets and rock, the space is blocked in different ways. Wooden or metal timbering, tow, rags and other materials are used for the purpose. Work on the isolation of support space occupy a lot of time and are very labour intensive.

A few types of pneumatic timbering systems (PBT) were developed by our team and passed wide industrial verification under various conditions of exploitation. They were used during building of underground constructions in such towns as Dnepropetrovsk, Moscow, Saint Petersburg, Protvino, Pavlograde (Buzylo 1987), (Andronov 1991).

During construction of distilling tunnels of Dnepropetrovsk underground passage, which was done by a drilling and blasting method in granites with the strength coefficient of f = 7–18 on the scale of prof. Protodyakonov, the shells used were of d = 0,4 m and length from 1,5 to 2,2 m with flat butt ends.

After editing of ring of tubings soft shells were laid on it and filled with compressed air. The idea of the PBT setting is presented on the figure 1.

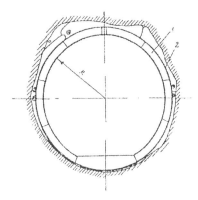

Figure 1. PBT setting during development of tunnels of subway in Dnipropetrovsk.
1 – tubing ring; 2 – soft shells PBT

Application of the PBT provides rapid and impermeable sealing of gaps, substantially lowers labour intensity of construction (up to 5-6 times), fully eliminates timber.

84

During building of the "Sestroretskaya (Saint Petersburg) station" the stationary timbering was tested. PBT was fastened onto the movable metal timbering. After the movement of the metal timbering, compressed air was applied into PBT under 0,03 MPa pressure, expanding the shell between the butt end of the metal timbering and ring of support and fully blocked the crack.

The PBT tests conducted under various conditions, including in excavation conducted by drilling and blasting method, showed its capacity, reliability, comfort of use. The time of assembling-dismantling works was smaller than in the traditionally used method. In addition timber is saved. It follows to suppose with a large degree of confidence, that PBT can be effectively used practically in all terms of conduct of support works.

During building of underground passages in weak rocks distilling tunnels are conducted, as a rule, by shield aggregates. However, workers are required to produce the expensive editing of shield aggregate and conduct short excavations with the help of pneumatic hammers. Thus the section of excavation transversal by beams is divided into tiers, which are developed from top to bottom. Timber support is traditionally used to prevent failure of roof and face rock. Timbering of the face of the excavation does not hinder development of mining pressure and resultant rock movement because of only partial contact between rock and support due to unevenness of the latter. As a matter of fact, the used temporary support is passive and is only removed layer by layer from an array rock shield working space of face from the hit in it already.

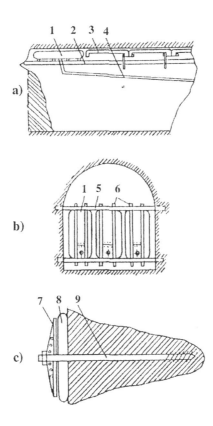

Figure 2. Temporary pneumatic supports preparatory excavations.
a – support of roof; b – support of face; c – in-bolt-pneumatic support.
1 – pneumatic support; 2 – sliding pipes; 3 – tubing support; 4 – air piping;
5 – fusillade; 6 – holding stake; 7 – face plate; 8 – pneumatic shells; 9 – bolt

The temporary pneumatic supports developed by us are active, that is having necessary initial hold and make high-quality alterations in the state of the surrounding solid. These supports hinder the exfoliation of rock, promoting safety of driving the excavation. A few constructions of such supports passed industrial verification during building of underground passages in Saint Petersburg and Kiev (Petrenko 1991).

When supporting roof by sliding pipes, which were suspended in collapsible on staples with rollers, pneumatic shells (fig. 2a) filled by compressed air were laid. Moving of sliding sander took place easily, and construction on the whole turned out as capable of working.

Temporary pneumatic support face was tested in excavation, which had a vaulted form with the area of face 20 m². For support of face a few soft shells in pillow form, which blocked all central part, were used (fig. 2b). Before development of rock every section was taken off in sequence, and reset after ending of coulisse. The achieved time of setting and cut support was on the average 3,5 minutes, that is approximately 10–15 times less than with the existent technology.

The tests showed that support constructions are capable of working and provide higher labour productivity and work safety.

Replacement of timber support by pneumatic in principle does not change the cost of the technology of conducting of excavation. The processes of laborious installation of crossbeams, their movement and installation, were labour intensive.

The newly developed support – pneumatic bolt (fig. 2c) generally eliminates traditional crossbeams.

One support section consists of a bolt, a soft shell and a face plate. Support principle of work consists in the following. A doughnut-shaped soft shell is installed with each bolt and held in place by a face plate. The applied compressed air expands the shell between the rock and the face plate, providing pressure onto the rockface and thus eliminating its breaking. Bolt-pneumatic support passed industrial verification during building of underground passage in Kiev (Buzilo 2004).

CONCLUSION

The new technical means of mechanization of processes of underground mining works described in the article – the pneumatic devices, and the positive results of their industrial verification prove the potential of their use when conducting mining works underground.

REFERENCES

Pneumatic Constructions In Mining. Kyiv – Donetsk. Higher Scholl 1983, p. 152.
Nekrasovsky J.E., Stepanovich G.J., Kazarov G.G.:.Exploitation of Pneumatic Support and Its Installation in Wall Space. Coal of Ukraine 1975, No. 9.
Khmarsky V.V., Andronov A.A.: Non-Static Pneumatic Constructions. Mining Journal 1982, No. 7, No. 5, p. 27–28.
Buzylo V.I., Ivanova I.V.: Pneumatic Support in Construction of Leningrad Metro. Metro constructing, Moscow 1987, No. 1, p. 13–14.
Andronov A.A., Buzylo V.I., Kolupajev O.V.: Pneumatic Face Timbering at Construction of Underground Constructions. Underground and Mine Construction. Moscow 1991, No. 8, p. 19–21.
Petrenko V.I., Rahutin V.I.: The New Hardware and Technological Processes at Constructions of Underground Buildings. Proceedings of the World Tunnel Congress '98 on Tunnels and Metropolises. Sao Paulo, Brazil 25-30 April 1998. A.A. Balkema/Rotterdam/Brookfield/ 1998, p. 361–364.
Buzilo Vladimir I., Rahutin Vladimir S.: Pneumatic Support – Technological Means for Mechanization of Underground Mining Processes. Ukrainsko-Polskie Forum Gornicze. Materialy Forum 2004. Dniepropietrowskij Narodowu Uniwersytet Gorniczy, p. 457–460.

International Mining Forum 2005, Sobczyk & Kicki (eds) © 2005 Taylor & Francis Group, London, ISBN 0415 375525

Development of Basic Criteria and Principles for Creation of Environment Friendly Technologies of Complex Development of Ore Deposit of Strategically Important Mineral Raw Material and Effective Technologies of Profitable Development of Poor and Off-Balanced Ores

A.E. Vorobiev
Peoples' Friendship University of Russia. Moscow, Russia

V.S. Pobyvanec
Peoples' Friendship University of Russia. Moscow, Russia

T.V. Chekushina
Russian Academy of Sciences
Institute for Complex Exploitation of Mineral Resources. Moscow, Russia

The transition to an extensive use of mineral resources during the deterioration of their quality assumes the development of essentially new strategy of their assimilation.

Absence of precise principles of mine design techniques in conditions of lack of detailed exploration of mineral deposit can become the reason for making inappropriate and even inconsistent decisions. The problems of creation of optimum solution at designing the basic technical processes and auxiliary systems of a mining enterprise differ by the complexity which accrues with increase in number of variants, time taken for making a decision, number of estimation aspects and number of the organizations which are affected by the decision.

To develop a new approach (principle) for the construction of effective technologies of subsoil use in conditions incompletely reconnoitred on compound structures deposits, it is preliminary necessary to analyse in details the list of the processes which are carried out for mining works.

The analysis of the specified processes shows, that the basic element influencing their basic parameters, is the type of used energy (tab. 1, fig. 1), i.e. at presence of a stronger energy source capacity of the device used in mining works has changed, but technology basic principles remained constant all this time.

Hence, the system analysis serves as a method allowing rationally using subjective judgment in the decision of compound structured problems. And criteria substantiation of efficiency is the major component in development of design decisions because the criteria in essence define the subsequent technology.

The analysis of mining enterprise structure, and also hierarchical subordination of designing problems assumes the presence of the proved approach to the problem of decomposition of this system by its division into subsystems and drawing up of sequence decision and optimisation of interconnected problems.

In these conditions rather important problem is the substantiation of criteria system allowing for each considered level of design problems, and also for each subsystem element to choose such characteristics and parameters, which would provide high system effectiveness as a whole. Thus it is necessary to be guided by an optimality (Trubetskoj, Krasnjanksij, Hronin) principle, which says: if the subsystems and objects compounding them for each level are optimum by the criteria, corresponding to a higher level system than all system is optimum.

Table 1. Communication of the basic mining processes with the types of used energy

PARAMETERS	BASIC PROCESSES OF TECHNOLOGY					
	Physical			Physical and chemical	Biochemical	Radiating chemical
Type of used energy (forces)	Gravitational	Mechanical	Thermal	Electrochemical	Microbiological	Radiating disintegration
Basic application	Moving of mountain weight	Destruction of mountain weight		Change of chemical and mineral structure of extracted ores		Destruction of mountain weight
				Branch of useful component from empty rocks		
Estimation of efficiency	0,5			0,3	0,6	0,09

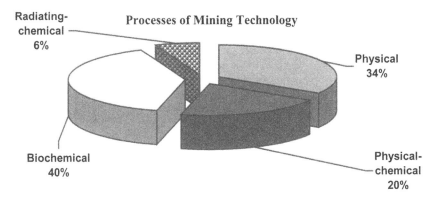

Figure 1. Efficiency estimation of basic mining processes

Generally, efficiency criterion represents an estimation of quality in performing its system functions. From establishment of decision efficiency as parities of expected useful effect (result) of the decision and necessary resources for it, it follows that efficiency criterion being a criterion of this correlation, should reflect it. From the resultant definition, the useful effect represents something else, as a degree of achievement of the established purpose. The resource is a set of potential elements of mining enterprise, which is necessary to create, support and own, for its functionality.

Criterion parameters should contain the factors of useful effect and expenses of resources. So, the generalizing condition consists in that it is more preferable to take a decision, which demands minimum expenses for resources, it is not permitted to consider as efficiency criterion, thus the results achievable at least expenses for resources are not accepted. The same way it is not enough to tell that it is desirable to find a solution, which gives the greatest useful effect. As a rule, there are always some variants of achievement of the same purpose, and more precisely, the preferable variant should be chosen with the help of the already accepted efficiency criterion.

The technology, used in mining enterprise, is necessary to present and to describe with mathematical model, which work will be characterized by a number of initial and final parameters - criteria (tab. 2). The construction of various communications inside the model and the description of their functioning dependences are described by the principles of model functioning.

Mathematical models can be divided into three interconnected classes: describing models, optimisation models and probable models.

Describing models – are the determined models, submitted in the form of the equations or inequalities, describing system behaviour. An example can be the differential equations of movement or restriction in the model of expenses and release.

Optimisation models contain expression, which should be maximized or minimized at certain restrictions. These expressions can be submitted in an algebraic or integrated form, in any other standard form, where are algebraic operations, integration or differentiation.

Table 2. Criteria of effective assimilation of reserve and off-balanced ores

CRITERIA	FACTORS	BASIC PARAMETERS (FACTORS)	OFFERED PARAMETERS (FACTORS)	ANALYTICAL AND EXPERIMENTAL DEPENDENCES
ECONOMIC	MINIMUM OF EXPENSES	$K + T_n \cdot C \to$ min. K – capital investments; T_n – normative time of recovery of outlay; S – the cost price.	FACTOR OF CAPITAL INVESTMENTS	
	MINIMUM OF EXPLOITATION ON PERIOD	$E_g = E_p + E_s$ – annual exploitation expenses; E_p – variable expenses; E_s – conditional-constant expenses.	FACTOR OF EXPLOITATION ON PERIOD	
	MAXIMAL PROFIT (PROFITABILITY)	$Re_f \geq Re_{min}$ – – profitability of the enterprise at volume of processing of mineral raw material Q_f; Re_{min} – the minimal level of profitability	PROFIT FACTOR	

89

Table 2 cont.

CRITERIA	FACTORS	BASIC PARAMETERS (FACTORS)	OFFERED PARAMETERS (FACTORS)	ANALYTICAL AND EXPERIMENTAL DEPENDENCES
TECHNOLOGICAL	THE MAXIMAL PRODUCTIVITY	$A_0 = B \cdot k_n / T_0 \cdot k_\kappa$ – – optimum annual production rate; T_o – economically optimum productivity of deposit processing. $B \cdot k_n$ – stocks taken from ores.	OPERATING RATIO OF THE EQUIPMENT	
	MAXIMAL EXTRACTION OF MINERAL RAW MATERIAL	$Ki = 1 - k_n = 1 - P/B$ factor of extraction of useful minerals; k_n – losses factor; P – size of losses; B – size of exhausted balanced reserves.	FACTOR OF EXTRACTION OF MINERAL RAW MATERIAL	
TECHNOLOGICAL	MINIMAL ORE TRANSFORMATION	$K_{s.} = V_B./V_\kappa.$ – – medium stripping factor; V_B – volume of stripping rocks; $V_\kappa.$ – reserves of useful mineral.	FACTOR OF ORE TRANSFORMATION	
		$K_k = f(K_f)$ – factor of complex processing of mineral raw material; $K_k = -0,001Q_f + 2,718;$ Q_f – quantity of processed mineral raw material, m^3/h.		

Table 2 cont.

CRITERIA	FACTORS	BASIC PARAMETERS (FACTORS)	OFFERED PARAMETERS (FACTORS)	ANALYTICAL AND EXPERIMENTAL DEPENDENCES
ECOLOGICAL	MINIMAL INFLUENCE ON THE ENVIRONMENT	$K_{ek} = -2{,}499\,K_k + 9{,}935;$ $K_{ek} = 0{,}004\,Q_f + 2{,}508;$ $K_{ek} = Q_f/Q_{ek}$ – factor of ecological processing; Q_f – quantity of processed mineral raw material, m^3/h; Q_{ek} – intensity of pollution with its ecological noxiousness, m^3/h.	FACTOR OF LIMIT ACCEPTED EMISSIONS	Factor of limit-accepted emissions $y = 0{,}1183e^{0{,}2508x}$ Quantity of processed mineral raw material m/h
	MAXIMAL USE OF MINERAL RAW MATERIAL	$P = B - B_s$ – losses size; B – size of extinguished balanced reserves; B_S – quantity of extraction of useful mineral; $K_n = D \cdot a/B \cdot S$ – factor of extraction of useful mineral from ores; D – total quantity of extracted ore; $s;\ a$ – maintenance of useful mineral in balanced reserves; in the extracted useful mineral.	FACTOR OF EXTRACTION OF USEFUL MINERAL	Maximal use of raw material $y = 14{,}533e^{-0{,}2508x}$ Quantity of extracted useful material.
	MAXIMAL PROTECTION OF PEOPLE ENVIRONMENT	Ecology factor $K_{эк} = Q_f/\sum Q_j\,(1+\xi);$ Q_f – quantity of processed mineral raw material, m^3/h; Q_j – intensity of pollution, j-type, m^3/h; ξ – Factor of ecological noxiousness of pollution.	FACTOR OF EXPENSES FOR UTILIZATION OF WASTE	Ecology factor $y = -3{,}6359Ln(x) + 10{,}172$ Intensity of pollutions

Probable models are such models, which also are expressed in equations and inequalities form, but have a probable sense (for example, it can be about mathematical expectations). In connection with that, decisions theory, being a branch of optimisation, is engaged in the maximization of utility average value, also within the framework of optimisation there are probable expressions and restrictions.

At the same time it is necessary to note, that the applicability of research methods in mining technologies is defined by the possibility to find precisely the criterion (purpose), and also to construct the formal model, expressing the communications between the criterion, variable and existing trough restrictions, to receive sufficient quantitative information, allowing to make a reasonable definition of parameters. In practice, it is more possible to perform these conditions after the solving of the problems from the bottom of hierarchical levels. However, significant successes in the application of operations research methods have caused the use of similar models in essentially different situations when the type of dependences between the variables, included in the formula, is not objectively determined and it is clear only, that one of them increase the value of the project, and the other reduce it. In the models, corresponding to such situations, it is reflected only by subjective confidence that selection of the projects should be carried out on the basis of the offered dependence.

At estimation of mining technologies it is preferable to use quantitative criteria because they provide, on the one hand, high objectivity regarding the estimation of condition of investigated system, and on the other hand, can be used at the establishment of actions efficiency, directed to transform the initial system.

The described process of elaboration of effective mining technology is represented in figure 2. On the scheme, continuous arrows represent the sequence of stages, and dotted arrows – the feedback relations, consisting in checking the solutions on stability, and also in change of the criterion.

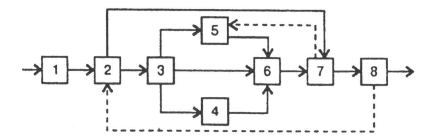

Figure 2. The block diagram of elaboration process of solution (Trubetskoj, Krasnjanksij, Hronin).
1 – establishment of the problem; 2 – choosing the criterion of efficiency; 3 – development of research model; 4 – establishment of the area of possible technical solutions and their characteristics; 5 – choosing the characteristics of external conditions; 6 – performance of calculations on the model; 7 – choosing of the optimum solution variant and check on its stability; 8 – the analysis of the results and development of recommendations

Usually the basic system characteristics are efficiency of their functioning and cost of production, i.e. at designing aspires to reach two purposes: increase of efficiency and reduction of system cost price. In this case it is used the method of technical and economic analysis of the systems, based on comparison between expenses and efficiencies.

For what, as a rule, is applied the criteria of economic character. However the spectrum of possible formulations of such criteria is rather wide: at the establishment of criterion can be used such economic characteristics, as total capital expenses, costs in time unit, net profit in time unit or production unit, incomes from investments, relation between expenses and profit, etc.

Besides this, at the estimation of technology of development of minerals deposits it is necessary to take into account also the criteria which are based on technological factors, for example, when it is required to maximize the rates of processing (to minimize the terms processing the deposit), to minimize the quantity of consumed energy, etc.

Mining industry is connected to high-risk level and traumatism. From here, the estimation of technology of deposit development of useful mineral by criteria of reliability and the maximal safety of works is necessary.

At the decision of practical problems of ecological estimation we should give an active character, which would promote the transition to efficient technologies and their purposeful improvement, since the basic possibility to reduce the level of emissions and waste in mining industry should be incorporated in the perfection of technology of extraction and processing of mineral raw material.

Therefore it is expedient to use as criteria of technology estimation the parity of actual and normative emissions for each substance.

However, the application of this criteria group is rather inconvenient due to two reasons.

The first reason is the significant quantity of different substances emitted by mining enterprises. So, stationary sources of technological-technical base of mining enterprises emit up to 50 and more types of polluting substances, moreover, as a rule, only 30–40 substances are permitted.

The second reason is the difference of chemical and biological properties of considered substances. In particular, all substances are subdivided into 4 classes of danger according to their properties and influence on the environment. However inside the class, each substance has its individual characteristics. Therefore, for the basic polluting-substances have been established some aggressiveness parameters, which determine the action of each substance on alive organisms and the probability of the accumulation of given substance in the environment and food circuits. Thus it was accepted, that $A = 1,0$ for carbon oxide. The use of the factor A, as quantitative criterion, allows to proceed from absolute values of emissions of 40 substances to one parameter of the mentioned emission which can comprise all spectrum of considered substances, i.e. to proceed to one criterion M_P.

However, this criterion also does not totally reflect the physical essence of the investigated processes and the degree of accomplishment of the established purpose.

In a large measure, we will approach to the reflection of the purposes as criterion, the relation of actual weight of the resulted emissions, factor of normative or permitted weight for the same period.

Because of the normative and sanctions regarding the pollution are established depending on the quantity of various substances, which have a negative influence on the environment and people, and also because of environmental contamination, the parameter of ecological safety of mining enterprise reflects better now a day the system technological processes and the degree of system influence on the environment and people.

Ecological damage develops from:
– additional social expenses in connection with environmental changes;
– expenses for the return of the environment in a former condition;
– additional expenses of future society in the connection with irrevocable withdrawal of deficit natural resources.

For the estimation of environmental damage there are used the following basis parameters:
– expenses necessary to reduce the pollution;
– expenses for environment recovery;
– market price;
– additional expenses registered because of change of environmental quality;
– expenses for compensation of risk for people health;
– expenses for additional natural resource for the dilution of thrown flow up to safe concentration of polluting substance.

More than 90% of all damage, which results from the placement of waste products, is connected to the processing of the useful mineral. Thus 60% of damage is connected to warehousing waste products of processing.

From the above-stated, it follows that the solution, suitable for the application of optimisation of mining technologies, unites the characteristic measure, a multitude of independent variables, with the model, which reflects the interrelation of variables. The primary goal of construction of the mathematical model describing technology of the mining enterprise, optimum "coordination" between principles of construction of technologies and criteria of an effective utilization of mineral raw material is.

As it can be seen, the estimation of offered technology is necessary to be done by several criteria. At the same time, only one criterion can be used for the definition of the optimum, because it is impossible to provide simultaneously, for example, the minimum of expenses, the maximum of reliability, the minimum of consumed energy, etc. In conditions when it is impossible to reach a simultaneous improvement of the solution by all criteria, the search for the best variant is reduced to finding of the most favourable compromise solution. Its choice from many criteria is connected to necessity to find the solution for replacements problems, i.e. the problems of comparison on preferability of losses by one criterion, which was better than another. To reach this it is possible with the help of different methods, which depend on the requirements of the final decision, character of a specific target of acceptance of the last, completeness of the information and of some other factors. In this case, any of criteria gets out as a characteristic measure (primary criterion). Other criteria are secondary and are used as limits of optimisation in corresponding ranges of change: from minimal up to maximal favourable value.

But mining technology (the preferable variant) can be also chosen by one criterion.

The choice of the most preferable variant is often made by using the criteria of Vald, Savedge, Gurvits, Laplas.

1. Vald's criterion – the greatest caution (maxim criterion). With its help it is chosen the best of the worst possible variants:

$$U_B^* = \max_i \min_j U_{ij} \qquad (1)$$

where i – line index (variant of solution); j – column index (external conditions).

At the application of this criterion, it is find the minimal parameter value of the solution U_{ij} for each variant, in this case corresponding to the worst state of external conditions, i.e. min a_{ij}. Further, from all possible variants, is chosen the variant which minimal parameter value of solution is the maximum. Vald's criterion is pessimistic, when it is applied, are guided by the worst state of external conditions.

2. Savedge's criterion represents the diversity of pessimistic approaches, when choosing the variant of problem solution. When it is applied it has to be find a minimal risk value at the most unfavourable state of external conditions:

$$U_C^* = \max_i \min_j r_{ij} \qquad (2)$$

With this purpose for each variant of external condition Π_j it is calculated the maximal parameter value U_{ij}. Then it is made a matrix of risks, and for that from the maximal received values U_{ij} are subtracted the sizes of this parameter for all variants of problem solving and in external conditions. The received results are included in the matrix of risks. In an additional column of this table it is shown the maximal risk value (loss) for each variant of solutions.

Further for each variant of solution it is find the maximal risk value from matrix of risks, and then it is chosen the minimal value from them.

3. Gurvits' criterion is combined, compounded from both optimistic and pessimistic approaches. When it is applied there are used some intermediate characteristics of external conditions, not the best, but also not the worst. The strategy, which is considered the most preferable, is:

$$U_\Gamma^* = k \min_i U_{ij} + (1-k) \max_j U_{ij} \to \max \qquad (3)$$

where $k = 0\text{--}1$ – represents the factor which characterizes the share of pessimism and optimism.

The factor k is chosen by subjective reasons: the must difficult is the situation and the more necessary to be insured; the factor k is closer to "one". When $k = 1$, Gurvits' criterion will be transformed into Vald's criterion.

4. Laplas' criterion is based on the principle of insufficient substantiation. Since the probabilities of situation are unknown, the necessary information for the conclusion that they are different is absent. All variants of external conditions are considered as equiprobable and for each alternative A_i it is determined the parameter:

$$U_{i.cp.} = \frac{1}{m} \sum_{j=1}^{m} U_{ij} \qquad (4)$$

Further it is chosen:

$$U_Л^* = \max_i U_{i.cp.} \qquad (5)$$

Usually there is no reason to prefer to any of resulted criteria. Therefore in the conditions, when there is no obvious domination of one alternative above another, the complexity of criterion choice becomes equivalent to the complexity of a direct choice of one of alternatives.

After the selection of the most effective criterion, in elaboration of leading basis of mining enterprises designing it is possible distinguish two different approaches (Trubetskoj, Krasnjanksij, Hronin).

One of them is used for the solving of project basic solutions of known technical solutions, used before in similar situations, i.e. has evolutionary character.

The basic solution consists in a more exact verification of large number of the factors, which provide the achievement of high parameters of known technical solutions. In these projects it is not possible to receive a qualitative growth of the basic technical and economic parameters in comparison with those already achieved.

Another approach is focused on the elaboration of essentially new technical solutions by means of new ideas and physical principles, modern materials and devices from other science areas and techniques. It is reached due to wide and deep analysis of the entire environment. It is obvious, that this approach is more universal and is perspective. At the same time we may observe, and this only conditional enough, because at their elaboration a significant number of technical solutions are taken on the basis of the last experience, i.e. use the principle of designing on prototype.

For designing of each mining enterprise there is no necessity for a detailed independent study of all environments. Such work can be replaced by a profound study of various informational authentic sources. However it is incorrect to exclude completely the procedure of environment analysis, because it leads to loss of such approach to the decision of problems.

During the further elaboration of methodical approaches of construction of highly effective mining technologies we have developed and offered to use the principle of flexible designing of mining enterprises (Vorobev, Balyhin, Karginov, Chekushina 2004). The given principle represent a list of various criteria, allowing to carry out the transition to efficient technologies (because in this technology are incorporated the basic opportunities to decrease the level of emissions and waste), and to their effective change (when changing mining-geological, technical-technological and economic conditions).

Thus, the principle of flexible designing of mining enterprises represents a system of dynamic designing technologies, modified depending on changes of:
- mining-geological parameters of minerals deposits (width of ore body, dip and depth, grade of useful component, physical-mechanical properties of ores and rocks, etc.;
- technical-technological parameters (productivity, reliability, etc., changing in time);
- economic characteristics (the cost of extraction and sale price of useful mineral).

Here is defined the specified principle on the basis of dynamic designing factor:

$$K_{\text{designing factor}} = \frac{K_{g-g} \cdot K_{t-t} \cdot K_e}{N} \qquad (6)$$

where K_{g-g} − influence factor of mining-geological parameters; K_{t-t} − influence factor of technical-technological parameters; K_e − influence factor of economic characteristics; N − quantity of technological changes of initial project in time unit.

Table 3. Estimation of basic factors, which provide an effective functioning of flexible designing principle

BASIC FACTORS	DETERMINING PARAMETERS	PARAMETERS		ESTIMATED POINTS
K_{g-g} − influence factor of mining-geological parameters	Width of ore body [m]	<5		1
		5−10		2
		10−30		3
		30−100		4
		100−500		5
	Depth [m]	<50		5
		50−100		4
		100−500		3
		500−1000		2
		>1000		1
	Angle of dip	Steeply inclined >25°		1
		Shallow dipping 5−25°		2
		Horizontal <5°		3
	Resources	Preliminary exploration (C₂)		1
		Off-balanced (A+B+C₁+C₂)		2
		Balanced (A+B+C₁)		3
K_{g-g} − influence factor of mining-geological parameters	Properties of ores and rocks	Competency	High	1
			Medium	2
			Weak	3
		Strength	Friable <6	4
			Weak-strong 6−12	3
			Medium 12−15	2
			Strong >15	1
		Fracturing	Monolith	1
			Coarse	2
			Medium	3
			Fine	4

Table 3 cont.

BASIC FACTORS	DETERMINING PARAMETERS	PARAMETERS		ESTIMATED POINTS
		Drilling	Easy drilling	3
			Medium drilling	2
			Hard drilling	1
		Modulus of elasticity [Pa]	$<5\cdot10^9$	4
			$(5-10)\cdot10^{10}$	3
			$10\cdot10^{10}-1,5\cdot10^{11}$	2
			$>1,5\cdot10^{11}$	1
		Density of rocks [kg/m³]	<1100	5
			$1100-2000$	4
			$2000-3000$	3
			$3000-4700$	2
			>4700	1
		Poisson ratio	$<0,15$	5
			$0,15-0,20$	4
			$0,20-0,30$	3
			$0,30-0,38$	2
			$>0,38$	1
		Compression strength [Pa]	$<1\cdot10^8$	5
			$2\cdot10^8$	4
			$3\cdot10^8$	3
			$4\cdot10^8$	2
			$5\cdot10^8$	1
		Tension strength [Pa]	$<0,5\cdot10^6$	5
			$0,5\cdot10^6$	4
			$1\cdot10^6$	3
			$1\cdot10^7$	2
			$2\cdot10^7$	1
	Grade of useful component in ores [%]	Pb [%]	<10	1
			$10-20$	2
			$20-30$	3
			$30-40$	4
			>40	5
		Zn [%]	<4	1
			$4-15$	2
			$15-30$	3
			$30-45$	4
			>45	5
K_{g-g} – change factor of technical-technological parameters	Productivity	Lower than nominal data		1
		Corresponding to technical characteristics		2
		Application of technological and technical parameters at their maximum parameters		3
	Reliability	Long outages		1
		Short-term outages		2
		According to PPR (scheduled preventive works)		3
	Fraction [mm]	$3-5$		1
		$2-3$		2
		$1-2$		3
		$0,1-1$		4
		$<0,1$		5
	Lumpiness [mm]	>1000		1
		$500-1000$		2
		$100-500$		3
		$20-100$		4
		<20		5

Table 3 cont.

BASIC FACTORS	DETERMINING PARAMETERS	PARAMETERS	ESTIMATED POINTS
	Enrichment [%]	<30	1
		50−30	2
		70−50	3
		90−70	4
		90	5
	Floatability [%]	<20	1
		50−20	2
		>50	3
K_e − change factor of economical parameters	Cost price [%]	Unprofitable production <100	1
		Profitable − 110−150	2
		Super profitable − more than 50	3
	Efficiency of capital investments	Minimal − less expected	3
		Normative − corresponding to projects	2
		Excess − exceeding the projects	1

The estimation of basic factors and parameters, which determine the parameters and estimation points of separate elements of mining enterprises, built on the basis of a principle of flexible designing, are resulted in table 3.

REFERENCES

Vorobev A.E., Balyhin G.A., Karginov K.G., Chekushina T.V.: Principle of Flexible Designing of Modern Mining Enterprises // Bulletin of Russian University of Peoples' Friendship. Series of Engineering Researches, № 1, 2004, pages 45−50.

Trubetskoj K.N., Krasnjanksij G.L., Hronin V.V.: Designing of Careers. M.: Publishing House AGN. Volume 1, page 519.

International Mining Forum 2005, Sobczyk & Kicki (eds) © 2005 Taylor & Francis Group, London, ISBN 0415 375525

Study of Dust Pollution in Slate Processing Plants. Integration of the Results into a GIS

C. Ordóñez
University of Vigo. Department of Natural Resources and Environmental Engineering. Vigo, Spain

J. Taboada
University of Vigo. Department of Natural Resources and Environmental Engineering. Vigo, Spain

M. Araújo
University of Vigo. Department of Natural Resources and Environmental Engineering. Vigo, Spain

ABSTRACT

In the history of mining, especially of coal, the dust produced by minerals and rocks has proved highly noxious to humans, causing illnesses such as silicosis, which, until recently, was a prime cause of death among miners. However, although studies of these negative effects on humans have mostly focused on underground coal mining, this kind of dust is also produced in mineral and rock processing plants, such as those for producing slate and granite. Galicia (Northwest Spain) is one of the world's main producers of granite and slate.

We studied dust contamination in a number of Galician slate production plants, some of which have existed for a number of generations and others, which are more recent. Dust concentration measurements were made for four particle sizes (0.5 μm as the maximum diameter) at different points of the plants, but particularly in the areas where people worked. Other variables that positively or negatively affected dust levels were also taken into account, such as noise levels or the existence of dust extractors. These data were then used to devise, using multivariate statistical methods, an environmental quality index that represented the level of contamination for each sampled point as a single value. In this way the most problematic working locations could be identified, with a view to implementing measures designed to reduce plant contamination levels. The results were represented in contamination maps constructed using kriging interpolation and integrated in a GIS (Geographic Information System) together with information on plant machinery, the kind of work performed at each location, and the characteristics of a dust extractor installation designed to reduce operator exposure to the dust generated by the slate-working processes.

1. INTRODUCTION

Commercial roofing slate is elaborated from slabs extracted from quarries in slate production plants. The process consists of the following phases:

1. Initial exfoliation: The blocks of slate are broken up, according to schistose layering, into slabs (maximum 30 cm thick). This process is performed using pneumatic or hydraulic compressors and perforators, or manually using mallets and wedges.

2. Sawing: The resulting slabs are loaded onto wagons using an overhead crane or fork-lift truck. Next, they are cut with the use of automatic diamond-disk saws, into parallelopipedic shapes slightly larger than those required for the final sheets of slate. They are then transported in metal containers or on rollers to the final exfoliation area.

3. Final exfoliation: In this process the parallelopipedic slabs are reduced to commercial thicknesses (3–6 mm), either by workers skilled in using hammers and wedges, or (increasingly common) by machines.

4. Cutting: This is the final shaping of the exfoliated sheets of slate into commercial sizes using special cutters or automatic machines.

5. Grading and packaging: The slabs are sorted according to size, thickness and quality, and then packaged on pallets for delivery.

Working conditions and labour productivity in slate plants have improved considerably in recent years, due to a more rational organization of the plant activities, the introduction of a higher level of automation in the sawing, exfoliation and cutting processes, and the utilization of continuous transport methods (García-Guinea et al. 1997). Nonetheless, the processes described have a common denominator in that they produce a high level of dust. Satisfactory solutions to this pollution problem have yet to be found, as the levels of dust in slate yards very frequently exceed acceptable limits. Any study that has as its purpose the resolution of the problem, however, is complicated by a variety of conditioning factors.

2. ENVIRONMENTAL CHARACTERISTICS OF DUST

Atmospheric pollution in the mining industry is caused by different compounds which, depending on their physical state, can be classified as solid/liquid particles or gases/vapours. The solid contaminants, better known as dust, generally measure between 1 and 1,000 µm, and, depending on the origin, vary greatly in terms of chemical composition. Dust, which eventually settles due to gravity, is the principal cause of pollution in mining activities. It is harmful in the following ways:

– It aggravates humans and damages the environment.
– It may be toxic and, as such, harmful to health.
– It prematurely wears out the moving parts of machinery.

This paper will examine a particularly harmful dust known as pneumoconiotic dust, which lodges in the lungs and causes a reaction in the tissues that leads to formation of fibrosis. The proportion of the overall dust cloud likely to be retained by pulmonary alveoli is denominated breathable dust and the particle size most likely to be retained is the micro. For the dust associated with slate quarrying to cause pneumoconiosis, it has to contain free silica and be of breathable size. In chemical terms, the most important components of slate are 48–53% silica (SiO_2) and 19–30% aluminium (Al_2O_3).

To determine dust contamination risk, Spanish regulations take into account two parameters (González A. available online): concentrations of breathable dust in mg/m^3 and the percentage of free silica in dust (S). The maximum permitted value (MV) for dust concentration is MV = 25/S, and levels at no time should exceed 5 mg/m^3 (Miner 1995).

3. CONTAMINATION CONDITIONING FACTORS

Although only two parameters are used to define the problem of pneumoconiosis from a Spanish legislative perspective, many other variables, in fact, need to be taken into account in any approach to assessing the dust contamination problem. Severity varies, for example, according to seasons and depends on whether the dusts originates in the mine or quarry itself or in the atmosphere, etc.

(Taboada et al. 1998), in their evaluation of contamination from a range of sources, have classified working locations for a single type of slate plant according to three contamination groups – high, medium and low – irrespective of dust particle size.

In order to fully understand and correctly assess the dust contamination problem, other variables that contribute to a greater or lesser extent to the overall environmental impact of slate quarrying need to be evaluated, as follows:

1. Time of year, i.e. winter or summer.
2. Free silica content of the dust (%).
3. Concentrations of breathable dust (mg/m^3).
4. Degree of compliance with existing legislation.
5. Existence of anti-dust measures.
6. Whether effects are direct or indirect.
7. Occurrence in conjunction with other types of pollution e.g. noise.
8. Temporal aspects, i.e. whether contamination is temporary or permanent.
9. Spatial aspects, i.e. whether contamination is local or general.
10. Catchments area, i.e. whether near or far from the dust source.
11. Number of people affected.
12. Whether or not the dust source can be isolated from other installations.
13. Whether or not the dust production process is reversible.

4. DETERMINATION OF A POLLUTION INDEX

Based on values for the above variables assigned to each working location, the most/least serious cases of atmospheric dust-pollution (highest/lowest concentrations of dust in conjunction with the best/worst conditions) were identified. Two classifications for the extreme cases were established. Of the 41 working locations assessed, 10 were classified as more serious contamination cases and 6 as less serious contamination cases. The remaining 25 cases were classified as average contamination cases.

Using the 16 extreme cases, a contamination index was constructed by creating a linear combination of the variables that best explained the initial classification and applying the criterion of minimizing the relationship between internal variance (within each contamination group) and total variance. Given that this is the criterion normally used in discriminant analysis (Taboada et al. 2002), it is possible to use the results obtained by means of this multivariant technique, and especially the discriminant function. Table 1 illustrates the coefficients of the linear combination for the significant variables.

Once the index values were calculated for each sample, each element was assigned to a contamination level (which may or may not have coincided with the initial assignment). The greater the degree of concurrence between the index contamination prediction and reality, the greater the precision of the linear function as a contamination index. Our results are depicted in table 2, where the 100% success rate indicates the absolute reliability of the index; in other words, each sample was correctly assigned.

The importance attached to each of the variables in identifying atmospheric contamination and their weightings in the linear index was assessed by means of the correlation coefficient for each variable and the index itself (see table 3). The most important variables were dust levels (in mg/m^3) and the proximity to the dust source. Similar conclusions were arrived at using the Wilks' lambda coefficient, with the associated significance level indicating that these two variables were significantly different for the two contamination groups.

In order to validate the results obtained, a sample that had not been used to determine the function was used to test whether or not the results of the prediction and the classification were valid. A cross-validation procedure was used that consisted of calculating the discriminant function using all sample elements minus one, and then applying the resulting function to classifying the isolated case. This calculation was repeated for each sample element. Table 4, which summarizes the overall result for the cross-validation, shows a success rate of over 80%, which would indicate the efficiency of the proposed function as a contamination index.

Table 1. Standard values of linear combination coefficients

VARIABLE	COEFFICIENTS
Time of the year	−0.046
Working location	0.325
% SiO$_2$	0.505
mg/m^3	1.364
Nearby source	0.519
Numbers affected	0.668
Impossible to isolate	−0.140

Table 2. Summary of classification results

ACTUAL GROUP	NUMBER	PREDICTION GROUP	
	of cases	0	1
Low	6	6 100%	0 0%
High	10	0 0%	10 100%
Medium	25	11 44%	14 56%

Table 3. Correlation coefficients for the linear function and the environmental variables

VARIABLE	CORRELATION COEFFICIENT
Time of year	−0.034
Job	0.004
% SiO$_2$	0.057
mg/m^3	0.527
Nearby source	0.257
Numbers affected	0.211
Impossible to isolate	0.185

Table 4. Results of the cross-validation

		PREDICTION GROUP	
ACTUAL GROUP	NUMBER OF CASES	CROSS-VALIDATION	
		0	1
Low	6	5 83.3%	1 17.7%
High	10	2 20%	8 80%

5. GEOSTATISTICAL STUDY OF DUST POLLUTION

The proven efficacy of the discriminant function in classifying sample elements according to contamination levels means that these values can be taken as representative of atmospheric dust pollution levels. Thus, rather than taking into account the 13 initial variables listed in Section 3, it is sufficient to determine the index value to identify a level of pollution for each working location. The simplicity of having to handle just one variable to describe what occurs at a given time and place is one of the great advantages of a statistical study of the initial variables. Another important advantage is the possibility of carrying out spatial estimations for the index. The atmospheric pollution levels calculated at specific locations throughout the plant extrapolated to the entire site provide a realistic picture of the highest pollution areas. Applications of this kind offer valuable information for devising appropriate anti-dust measures.

The techniques used in the geostatistical study permit an analysis of the dependency structure of what are known as regionalized variables, i.e. variables defined for all points in space. See (Cressie 1993) or (Journel et al. 1993) for further details. Once spatial dependency for a regionalized variable has been modelled, we can use a set of sampling values to estimate points in space that were not included in the initial sample.

The starting point for our study was the 41 working locations identified above, which corresponded to a standard slate plant measuring 40×30 metres. For each working location, the variables that characterize the corresponding level of atmospheric pollution were assessed and the corresponding pollution index value was calculated using the discriminant function. Each value associated with a set of coordinates identify it specifically and exclusively with a single working location. In our case we have 41 values, distributed over the entire area of the slate plant, corresponding to a single regionalized variable (the contamination index). An experimental semivariogram was calculated with a view to determining and modelling the spatial behaviour of this regionalized variable. This function, dependent on the distance h between two working positions, assumes the values that are represented in figure 1 by means of points. Next we sought to obtain the theoretical semivariogram function that best fitted the cloud of points.

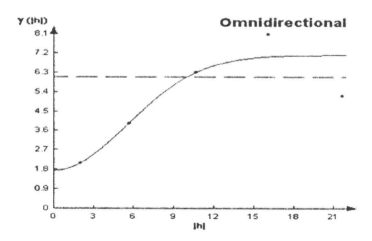

Figure 1. Theoretical and experimental semivariograms for the dust contaminination index

From among the different models that could be considered theoretical semivariograms, the best fit was found to be a structure of 3 nested semivariograms, as follows:

A nugget model with a value of 0.154.

A gaussian model with range 16.32 and sill 1.05.

A spherical model with range 17.52 and sill 0.264.

The semivariogram resulting from the sum of the three produced the function shown as a continuous line in figure 1.

This completes the description and analysis phase of the evolution of the index between work locations in the plant.

For the next phase – estimation of pollution values for the entire site – it was necessary to resolve a system of equations for each point for which an estimation was required, using an approach known as kriging. The theoretical semivariogram obtained as described above was used to calculate the coefficients.

6. INFORMATION HANDLING AND PROCESSING USING A GIS

Dust pollution was studied at four plants, classified as first generation (the oldest) to fourth generation (the most recent). Maps of the installations for each plant, together with those of dust pollution calculated using geostatistics, were used to construct a GIS that helps us to manage all the existing information, compare the different plants studied and design a ventilation system to reduce the pollution, especially in the working locations with a big quality index.

The GIS was developed as an independent Visual Basic application based on GeoMedia components.

The GIS lets the users make spatial queries (for example, to see where the saws are located in each plant) and attribute queries (for example, to a known value of the quality index in a specific point), or simply view the different maps previously imported, such as the following:

- A plan of the plant showing the different working areas, machinery and operator workstations (fig. 2).
- A map of total dust concentrations based on the data measured directly in the plant.
- A map of the statistical contamination index showing the plant contamination.

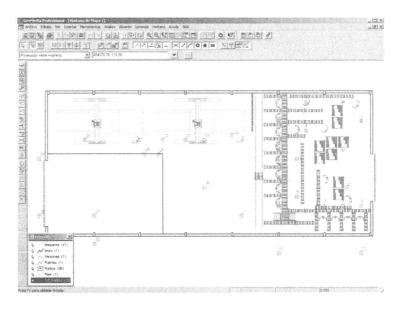

Figure 2. First-generation slate plant

The statistical contamination index that represents contamination caused by dust and other variables constitutes the most important result of this study. Atmospheric pollution levels calculated at specific locations in a plant were extrapolated to the entire plant, to offer a more realistic picture of the areas of highest pollution. This index, based on values ranging from 1 to 10, represents different degrees of pollution by means of closed polygons and shades of green. Applications of this kind offer valuable information from the perspective of developing appropriate anti-dust measures. The results obtained for each plant are explained and interpreted below.

- The sawing area in the first-generation slate plant is not physically separated from the final exfoliation, cutting and packaging area, so it is hardly surprising that the environmental contamination map (fig. 3) reveals the highest pollution levels to be in the sawing and packaging areas. The continual movement of wagons and a lack of cleanliness can explain the fact that the packaging area has a high level of contamination.
- The second-generation slate plant is divided into clearly differentiated zones for sawing, final exfoliation, cutting, and packaging. The highest values in the environmental contamination index (fig. 4) are in the sawing area (the result of the automatic diamond-disk saws), as also in the exfoliation and cutting areas, particularly in zones without dust extractors.
- The third-generation slate plant has three clearly distinguishable zones: two twin plants for the elaboration of slate, separated from each other by an area for final exfoliation, cutting and packaging. Walls that mark off different task areas further divide the elaboration areas. Surprisingly, significant differences were found between the twin plants (fig. 5), which can be explained by the existence of an extractor system in one of the two plants, but not in the other. As would be expected, the sawing area showed the highest levels of contamination.
- The fourth-generation slate plant is divided into two areas, with a sawing area that is entirely independent of the slate finishing zones. This would explain the great difference in index contamination levels at either side of the dividing wall, with the highest levels observed for the sawing area (fig. 6).

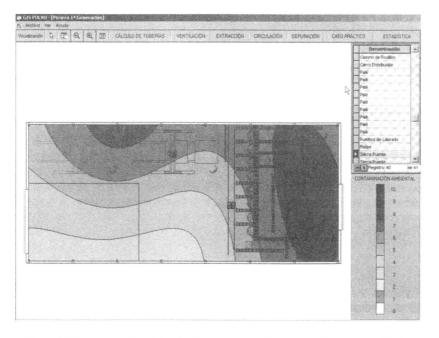

Figure 3. First-generation slate plant: environmental contamination statistical index

105

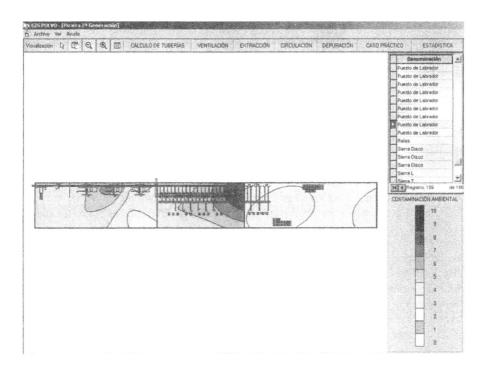

Figure 4. Second-generation slate plant: environmental contamination statistical index

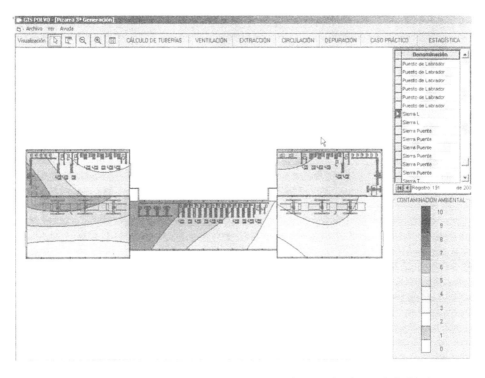

Figure 5. Third-generation slate plant: environmental contamination statistical index

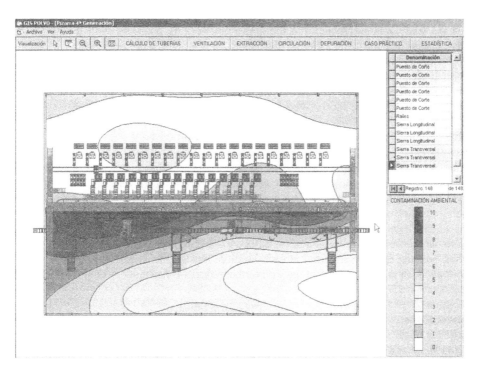

Figure 6. Fourth-generation slate plant: environmental contamination statistical index

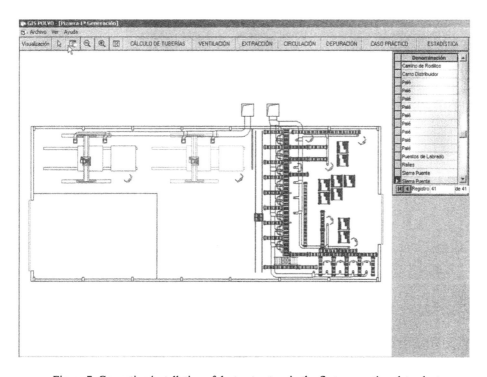

Figure 7. Corrective installation of dust extractors in the first-generation slate plant

Once the study was completed and the areas and working positions with the highest levels of contamination had been identified, a dust extraction system was designed to be installed at suitable locations in the four plants (first generation to fourth generation). This installation was incorporated as another layer in the GIS, so that it could provide information on the pipe diameters and airflows needed for each case (fig. 7).

8. CONCLUSIONS

Resolving the problem of high levels of dust pollution in slate plants required an assessment of the atmospheric factors that contributed to this contamination. These individual factors were measured at each working location with a view to creating an overall dust contamination index for each location. This index was then extrapolated to the entire plant with the use of geostatistical methods. Given that in terms of distribution, contamination tended to be greater in the central parts of the yards, the solution to the problem was relocation of the exfoliation and sawing processes to different parts of the plant, thus isolating the dust produced at each location. Finally, the installation of dust extractors at each working location in these areas of the yard further reduced concentrations of atmospheric dust.

REFERENCES

Cressie N. 1993: Statistics for Spatial Data. John Willey & Sons Inc. New York, USA.
García-Guinea J., Lombardero M., Roberts B. and Taboada J. 1997: Spanish Roofing Slate Deposits. Transactions of the Institute of Mining and Metallurgy, Section B, Vol. 106, B205–B214.
González A.: Control del polvo y del ruido en explotaciones de áridos: medidas de prevención. III Jornadas de Formación en Seguridad Minera en Canteras.Graveras y Plantas de Tratamiento de Árido. Available from http://concretonline.com/jsp/articulos/canteras2.jsp
Journel A.G. and Huijbregts Ch.J. 1993: Mining Geostatistics. Academic Press. New York, USA.
Miner 1995: Reglamento General de Normas Básicas de Seguridad Minera e Instrucciones Técnicas Complementarias. Centro de Publicaciones. Ministerio de Industria y Energía. Madrid.
Taboada J., Vaamonde A., Saavedra A., Argüelles A. 1998: Air Pollution in Slate Industry. Second European Conference on Geostatistics for Environmental Applications (GeoENV 98). Valencia (Spain), pp. 99–108.
Taboada J., Vaamonde A., Saavedra A., Ordóñez C. 2002: Geostatistical Study of the Feldspar Content and Quality of a Granite Deposit. Engineering Geology 65, pp: 285–292.

International Mining Forum 2005, Sobczyk & Kicki (eds) © 2005 Taylor & Francis Group, London, ISBN 0415 375525

Estimation of Environmental Impacts of Mining Technologies

Carsten Drebenstedt
Technische Universität Bergakademie Freiberg. Freiberg, Germany

Pierre Schmieder
Technische Universität Bergakademie Freiberg. Freiberg, Germany

1. INTRODUCTION

The close vicinity of mining companies to other sites that are used for commercial and residential purposes or to areas under conservation as well as the fact that the people become more and more sensitive to emissions have led to a negative attitude towards mining operations in the neighborhood. It is above all the use of drilling and blasting techniques that give rise to objections, in the course of which demands for more environmentally-friendly production methods have been voiced. It is in this context that production methods without the use of explosives are considered as an option. After all, the use of such production methods is the only alternative for some quarry operators. The question is what are the true benefits of these techniques and what is the relation between a possible improvement of the environmental effects on the one hand and the technological potential as well as the costs on the other.

It is intended to assess the mining techniques for solid rock in their entirety. This paper will therefore look at the aspects for assessing mining techniques in the context of their environmental impact and the costs.

2. PRELIMINARY REMARK

The "mining technique" comprises the "extraction system" and the actual "mining operations" as its two components. An "extraction system" refers to the chain of equipment necessary to exploit a deposit. In general, this chain of equipment is made up of partial operations, such as hewing and digging out the rock (= loosening), loading, hauling (conveying) and pre-crushing. The component "mining operations" is characterized by the direction of the mining progress (i.e. vertical or horizontal), the sequence of starting the individual slices and the mode in which the extraction system progresses within a slice (parallel mining, surface bench mining, side-to-side mining).

The term "extraction technique" refers to the way how the rock is loosened, which can be done either by drilling, by blasting or by a technique without the use of explosives. Figure 1 illustrates the extraction techniques schematically.

When talking about "environmentally-friendly mining techniques" one has not necessarily and automatically any mechanical extraction techniques without the use of explosives in mind, it is rather the aspects referred to in the definition (in fig. 2) which matter (Wehrsig; Drebenstedt 2002). In a narrower sense, this definition refers to all phases of the mining activities, beginning with the exploration, followed by the opening-up of the mine, the regular operations and eventually by the renaturalization/land reclamation, and looks also at the utilization or the consumption of operating equipment and consumables.

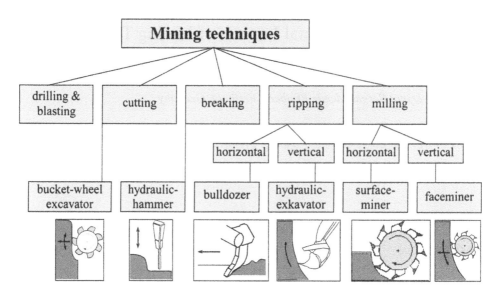

Figure 1. Extraction techniques for the mining of rock

In a broader sense, the definition in figure 2 includes additionally the phases of
– producing and disposing the operating equipment (loading, conveying and drilling equipment) as well as
– producing the consumables (energy, explosives, lubricants).
During these phases, both material and energy will be consumed and the environment will be affected by emissions into the atmosphere, by noise and by a destruction of the landscape.

The application of environmentally-friendly mining techniques means:

A.: avoiding or minimizing

> - **emissions into the atmosphere (exhaust gas, dust)**
> - **noise**
> - **vibration**
> - **the contamination of the ground and surface water (and thus maintaining the water quality) and**
> - **the creation of by-products (especially the share of non –marketable raw materials)**

B: reducing

> - **the consumption of consumables (fuel, power, explosives, lubricants)**
> - **the wear and tear**
> - **the temporary use of land (earlier reclamation/renaturalization)**
> - **the visual impairment of the landscape concerned and**
> - **the production losses (careful use of resources)**

Figure 2. Definition of an environmentally-friendly mining technique

3. METHODOLOGY FOR DETERMINING AN ENVIRONMENTALLY-FRIENDLY MINING TECHNIQUE

Figure 3 shows the flow chart for selecting an "environmentally-friendly mining technique". The annual production capacity (the output) of the open-cast mine to be assessed has been specified. Based on the deposit or the rock formation with its bonding or rock properties, the production capacity for each extraction technique will have to be established. The rock parameters, such as the pressure resistance (one-axial pressure resistance), the tensile and shearing strength, the modulus of elasticity and the bed rock parameters, such as the crevasse formation, the degree of withering and the formation of layers, all influence the loosening/breaking/cutting properties of the rock. In addition to that, the equipment weight and the loosening techniques are of the essence.

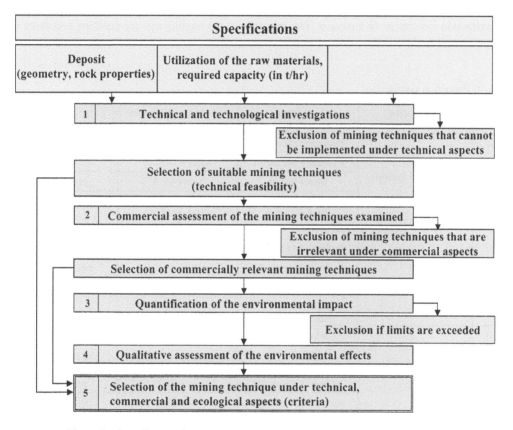

Figure 3. Flow diagram for selecting environmentally-friendly mining techniques

Further specifications, such as the utilization of the raw material and the resultant quality requirements as well as additional conditions imposed with regard to the environmental impact, such as the compliance with certain limits, will have to be observed when selecting suitable mining techniques. The calculation of the production capacity for each extraction technique alone has proved to be difficult so far, since there is no model available that provides details about the capacity for loosening the rock, when a certain extraction technique is employed (Niemann-Delius; Hennig 2004). However, there are individual analyses, although most of them only look at one single

111

extraction technique. This means that the pressure resistance in connection with the existence of parting planes (interfaces) is of major importance (Todzi 2000). Unfortunately, the capacity for loosening and loading the rock can only be determined to a certain extent by taking into account the rock or the rock formation with its relevant properties. The manufacturers of extraction equipment have their own methods to derive the loosening capacity of the machine concerned from the rock to be tackled. These methods are largely based on long-standing experience and cannot be applied ad lib to other machines. Certainty is eventually only gained on the basis of time and cost intensive onsite tests.

Numerous tests described in the literature as well as own practical experience make it possible to estimate the production capacity of a machine on the basis of the pressure resistance and a description of the interface structure, so that this methodology can be applied.

The technical/technological investigations will largely comprise the mining technique, which consists of the extraction system and the mining operations as its two components. Table 1 shows the extraction systems to be analyzed. The sequence of the hauling and pre-crushing processes has been reversed in some of the extraction systems.

Table 1. Extraction systems

	LOOSENING	LOADING	HAULING AND PRE-CRUSHING	
1.	drilling and blasting	hydraulic excavator	heavy truck	pre-crusher
2.		hydraulic excavator	mobile crusher	belt conveyer
3.		wheel loader	heavy truck	pre-crusher
4.		wheel loader (load & carry)	semi-mobile pre-crusher	belt conveyer
5.	bucket-wheel excavator		belt conveyer	pre-crusher
6.	hydraulic hammer	hydraulic excavator	heavy truck	pre-crusher
7.		hydraulic excavator	mobile crusher	belt conveyer
8.		wheel loader	heavy truck	pre-crusher
9.		wheel loader (load & carry)	semi-mobile pre-crusher	belt conveyer
10.	ripper	wheel loader	heavy truck	pre-crusher
11.		wheel loader (load & carry)	semi-mobile crusher	belt conveyer
12.		hydraulic excavator	heavy truck	pre-crusher
13.		hydraulic excavator	mobile crusher	belt conveyer
14.	hydraulic excavator		heavy truck	pre-crusher
15.			mobile crusher	belt conveyer
16.	surface miner	wheel loader	heavy truck	*
17.	surface miner		heavy truck	*
18.			belt conveyer	*
19.	face miner	wheel loader	heavy truck	*
20.	face miner		heavy truck	*
21.			belt conveyer	*

* no pre-crushing, as a low grain size will be generated during the loosening process

Mining techniques that cannot be implemented for technical reasons will be disregarded in all future considerations. Examples are soil class 7 and a low degree of withering, where a hydraulic excavator or a bucket-wheel excavator can hardly be employed for direct mining, given the current state-of-the-art. The extraction capacity is far to low or the wear and tear would be forbiddingly high. An extraction output of more than 300 000 t/a is possible indeed with a 115 t hydraulic excavator in direct mining operations, where the rock has a high pressure resistance and a low degree of withering (Schmieder, Bui Nam 2003).

The technically reasonable options will be examined under commercial aspects and on the basis of the methodology defined. This will include both static and dynamic cost analyses, with the static cost analysis examining the capital and operating costs. The dynamic cost analysis will be performed on the basis of the FMK method. Non-viable methods will be disregarded in all future examinations.

The commercial assessment will be followed, if possible, by a quantification of the environmental effects by way of an input-output analysis (fig. 4), during which the following environmental effects will be looked at:

- noise;
- dust;
- vibration;
- the CO_2 equivalent;
- the SO_2 equivalent;
- the cumulative energy input (KEA);
- the temporary use of land.

The vibration (Vibration velocities) caused by drilling and blasting operations will be calculated with generally valid equations. The value of the Vibration velocity from drilling and blasting will still have to be compared with the standard values stipulated in DIN 4140, part 3. The vibration occurring with all other techniques can only be estimated under qualitative aspects (Schmieder, Drebenstedt 2003).

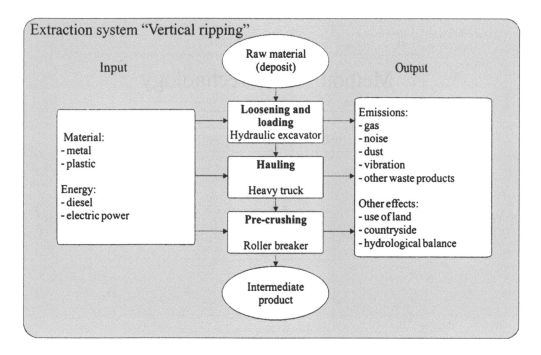

Figure 4. Input-Output analysis, using vertical ripping as an example

The mean sound radiation level is an indicator of how to assess the noise emission, while the quantities of dust generated can be calculated with the help of the methodology contained in VDI standard 3790, sheet 3. In order to do so, the emission factors of the processes in the extraction system will be established (Schmieder, Drebenstedt 2003).

The entire life process covering the generation of primary energy and raw materials right up to their utilization is recorded in the GEMIS database (Global Emissions Model of Integrated Systems), which also includes the auxiliary energy and the material input for the erection of plants as well as for the disposal processes. This database provides data for the diesel and power supply, the production of metal and plastic, which will be taken into account with regard to each and any extraction system considered (GEMIS 2000).

4. ASSESSMENT

The assessment of an environmentally-friendly mining technique should not only take the environmental effects into account, but also economic/commercial and social aspects. As a result of using machines and equipment and thus energy, emissions, such as CO_2, SO_2 and exhaust gas from the combustion engines, as well as noise, dust and vibration will be discharged or generated, respectively (fig. 5). The mode of operating the equipment (technology) will mainly impact the landscape (temporary use of land), the water balance and the opening-up of the site with the deposits. Apart from having an environmental impact, any mining technique will also incur costs (production costs).

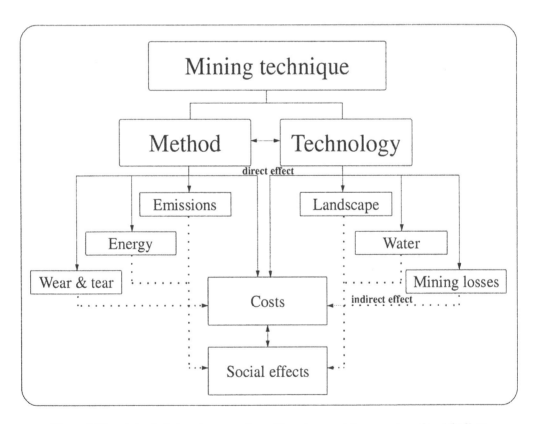

Figure 5. The mining technique in connection with environmental, economic and social effects

One will also have to consider, to which degree a mining technique gives priority to the social requirements under cost and environmental aspects, because an environmentally-friendly and cost-effective technique should also be socially acceptable.

The environmental, economic/commercial and social aspects are close interrelated, because the lower the environmental impact the cleaner and healthier the workplace, apart from the lower number of complaints from the neighborhood. Other social aspects include, but are not restricted to, are affordable raw materials, supply security, work places, secure employment, health protection and an environment that is worth living in.

If a reduction of the detrimental environmental effects is required or if the damage caused by the environmental impact of an open-cast mine needs to be rectified, further costs (external ones) will be incurred, apart from the production costs as such. An example is the recultivation of the post-mining land, which will increase the quality of life during and after the mining activities (social impact).

It is not possible at the moment to quantify all effects of extracting the rock or of the mining activities without incurring undue costs. This is why the peripheral conditions for the assessment have to be determined and why only production costs and the environmental effects are to be directly taken into account as specific assessment criteria for the time being.

First of all, the criteria of each technological alternative are summarized in a table. In order to be able to use these criteria in an evaluation matrix, they have to be scaled. This can be done by using an ordinary or a proportional scale. On an ordinary scale, the values of the same type of criteria are compared with each other, with the values being assigned points by taking into account the ranking among those criteria. In that case, the relative differences between the values of the criteria will be lost, which is a disadvantage and leads to a distortion of the information gained. On a proportional scale, the values of the same type of criteria will be scaled in a differentiating way (environmental effects), so that the content of the information can be largely retained. Taking the noise level as an example, it will have to be noted that its increase by a mere 3 dB means twice the intensity. Table 2 shows an evaluation matrix. The technical alternative A_i with the lowest environmental impact U_j gets the highest number of point p_{ij}, i.e. 100 points as an example. Fewer points have been calculated for the other alternatives in accordance with the distance from A_i within the criterion. The points p_{ij} awarded for the environmental impact of each mining technique will then be added up vertically at the bottom of the table. This sum will represent the environmental impact of the relevant mining technique expressed in points, with the largest figure pointing to the least environmental impact, which must be assessed as positive in the sense of environmental friendliness.

In order to differentiate further, additional corrective and weighting factors need to be introduced, which are shown in table 2.

The corrective factor k_{ij} takes the environmental effects of an environmental impact quantity U_j between the alternatives into account, thus taking care of the differences between the types of noise emissions or the types of vibration. The noise emission of a machine will have to be assessed differently than that of an explosion, to give just one example. Likewise, the vibration generated by a hydraulic hammer is different from that of a blast, both as far as time and intensity are concerned. The factor k_{ij} is thought to point out such differences, which will have to be verbally explained, if necessary.

The weighting factor g_j weighs the criteria U_j against each other. One will have to consider what impact local or global environmental effects have on an object that is to be protected. The fixture of this factor depends eventually on the preferences of the decision-making person and/or the factor can also be established by questioning the persons concerned. The latter requires extensive surveying methods which shall not be detailed in this paper. Instead, two basically different cases will be looked at: The first one refers to a mining company that is located far away from any residential areas or an object to be protected. Here, local environmental effects, such as noise, dust and vibration lose their importance. In the second case, the open-cast mine is close to a residential area, so that local environmental effects need to be considered, apart from the regional and global ones. These differences will be taken care of by the weighting factor g_j (fig. 6).

Table 2. Evaluation matrix

RANGE OF A CRITERION	CRITERION j	WEIGHTING	ALTERNATIVE A_i A_1	A_2	A_3	A_4	A_5
			k_{1j}	k_{2j}	k_{3j}	k_{4j}	k_{5j}
UE	U_1	g_1	p_{11}	p_{21}	p_{31}	p_{41}	p_{51}
	U_2	g_2	p_{12}	p_{22}	p_{32}	p_{42}	p_{52}
	U_3	g_3	p_{13}	p_{23}	p_{33}	p_{43}	p_{53}
$UE_{ges\ i} = \Sigma$			$\Sigma p_{1j}*g_j*k_{1j}$	$\Sigma p_{2j}*g_j*k_{2j}$	$\Sigma p_{3j}*g_j*k_{3j}$	$\Sigma p_{4j}*g_j*k_{4j}$	$\Sigma p_{5j}*g_j*k_{5j}$

UE – environmental impact; A – technological alternative (mining technique); i – variant of the mining technique (e.g.:1 – drilling and blasting, 2 – horizontal milling, ...); U – environmental impact; j – type of environmental impact (e.g.: 1 – noise, 2 – SO_2 equivalent, ...); p_{ij} – points awarded to mining technique A_i and to the environmental impact U_j (takes the differences of U with regard to A_i into account); k_{ij} – corrective factor for the type & intensity of the environmental impact, depending on A_i; g_j – weighting factor between the criteria U_j not depending on A_i.

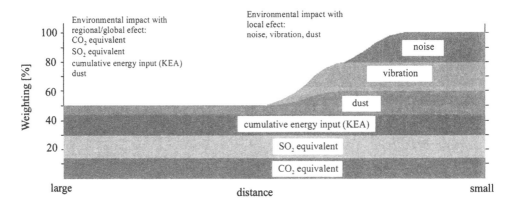

Figure 6. Weighting of environmental effects (diagram)

The following general equation for the overall environmental impact caused by a mining technique can be defined from table 2 and on the basis of the above passages:

$$UE_{ges\ i} = \sum_{j=1}^{n} p_{ij} \cdot g_j \cdot k_{ij} \qquad (1)$$

The result of this analysis will be a matrix, in which the environmental effects of each mining technique are quantified by a non-dimensional number. It is now possible to find out by assessing the matrix, which of the techniques have the least environmental impact in toto, but also with regard to each individual criterion.

The economic/commercial criteria will be additionally taken into account by the addend $W_{ges\ i}$ in equation 2. This addend will eventually have to be established in accordance with the schema as shown in table 2. It is also conceivable to observe the social advantages by way of introducing another addend $S_{ges\ i}$ into equation 2.

$$UEWS_{ges\ i} = UE_{ges\ i} + W_{ges\ i} + S_{ges\ i} \qquad (2)$$

The procedure explained above will make it possible to establish the most environmentally-friendly, commercial and/or socially acceptable mining technique.

5. APPLICATION

One example has been selected to examine the environmental friendliness of extracting rock by looking at the processes of loosening, loading and hauling the material. The rock was loaded with a wheel loader with a bucket of 5 m³, while it was shipped on a heavy truck with a loading capacity of 24 m³. The mining techniques of vertical ripping, horizontal milling and cutting did not require any loading equipment, because the loosening and loading processes were jointly carried out by the loosening equipment. When examining the cutting process with the bucket-wheel excavator, the extraction system was examined with the equipment combination of a bucket-wheel excavator, a mobile conveyer and a belt conveyer system. Both the face miner and the extraction system required power for the cutting operations, while all other extraction systems required diesel fuel.

It is assumed in the case of a fissured rock formation with a pressure resistance of 20 MPa that each extraction technique as listed in figure 1 will be applicable. The company's loosening capacity has been fixed at 300 000 t/a. The examinations have been carried out by taking environmental effects, such as dust, vibration, the CO_2 equivalent, the SO_2 equivalent and the cumulative energy input (KEA), into account. The costs (both operating and capital costs) have also been included in the assessment. The examination was based on inspections of open-cast mines for solid rock as well as on analyses made at the Freiberg Mining Academy (TU Bergakademie Freiberg) (Schmieder 2002–2004), (Schmieder 2001), (Schmieder, Bui Nam 2003).

Table 3 shows the quantity units for examining and establishing the environmental effects and the production costs.

Table 3. Environmental effects and costs of the mining technique (quantity units)

	Drilling and blasting	Horizontal milling (surface miner)	Horizontal ripping (dozer)	Hydraulic hammer	Bucket-wheel excavator	Vertical milling (TSM)	Vertical ripping (hydraulic excavator)
SO_2 equivalent [g/t of rock]	8.0	15.1	13.5	15.4	1.4	8.9	9.2
CO_2 equivalent [g/t of rock]	948.3	1.598.0	1.449.7	1.647.4	505.2	2.177.5	993.2
KEA [MJ/t of rock]	10.6	20.8	18.9	21.5	12.7	32.4	13.0
Sound energy level [dB(A)]	112.9	114.0	113.5	114.6	108.5	114.0	110.7
Dust [g/t]	120.0	190.0	370.0	205.0	400.0	900.0	160.0
Vibration	within the limit	low	low	less than drilling & blasting	low	low	low
Production costs [%]	104.8	159.3	149.9	143.5	121.9	207.9	100.0

As has already been implied in section 3, the vibration was mainly established by qualitative rather than numerical values. It must therefore be assumed that the vibration (intensity) during the blasting is very high as compared with the other mining technique, although the highest degree of vibration among all mining techniques without the use of explosives is caused by breaking. However, the vibration caused by breaking is still much lower than that during drilling and blasting operations.

The environmental effects have been recorded on differentiating scales. The two cases assumed, i.e. the open-cast mine far away from human settlements and the one close to a residential area, make it necessary to weigh the environmental effects differently (figure 6). Tables 4 and 5 as well as figure 7 show the effects of weighting the final results.

Table 4. Weighted environmental effects caused by the far-away open-cast mine

	Weighting	Drilling and blasting	Horizontal milling (surface miner)	Horizontal ripping (dozer)	Breaking (hydraulic hammer)	Cutting (bucket--wheel excavator)	Vertical milling (TSM)	Vertical ripping (hydraulic excavator)
SO_2 equivalent	15%	2.66	1.41	1.57	1.38	15.00	2.39	2.32
CO_2 equivalent	15%	7.99	4.74	5.23	4.60	15.00	3.48	7.63
KEA	15%	15.00	7.66	8.41	7.42	12.55	4.92	12.25
Dust	5%	5.00	3.16	1.62	2.94	1.50	0.67	3.75
Sound energy level	0%	0.00	0.00	0.00	0.00	0.00	0.00	0.00
Vibration	0%	0.00	0.00	0.00	0.00	0.00	0.00	0.00
Sum (ranking)	50%	30.65 (2)	16.97 (4)	16.83 (4)	16.34 (4)	44.05 (1)	11.46 (5)	25.95 (3)
Scaling	100	69.59	38.53	38.22	37.10	100.00	26.02	58.92

Table 5. Weighted environmental effects and costs caused by a close-by open-cast mine

	Weighting	Drilling and blasting	Horizontal milling (surface miner)	Horizontal ripping (dozer)	Breaking (hydraulic hammer)	Cutting (bucket--wheel excavator)	Vertical milling (TSM)	Vertical ripping (hydraulic excavator)
SO_2 equivalent	15%	2.66	1.41	1.57	1.38	15.00	2.39	2.32
CO_2 equivalent	15%	7.99	4.74	5.23	4.60	15.00	3.48	7.63
KEA	15%	15.00	7.66	8.41	7.42	12.55	4.92	12.25
Dust	15%	15.00	9.47	4.86	8.82	4.50	2.00	11.25
Sound energy level	20%	19.23	19.05	19.13	18.94	20.00	19.04	19.61
Vibration	20%	2.00	16.00	16.00	10.00	16.00	16.00	16.00
Sum (ranking)	100,00%	61.88 (3)	58.33 (4)	55.2 (5)	51.16 (6)	83.05 (1)	47.83 (7)	69.06 (2)
Scaling	100	74.51	70.24	66.48	61.61	100.00	57.60	83.16
Costs (scaled)	100	95,45	62,77	66,71	69,67	82,01	48,11	100,00

As has already been mentioned, the costs must not be disregarded when an assessment is made on the basis of the environmental friendliness. Like the environmental effects, the costs have also been subjected to a differentiating scaling (table 5), in which case the mining technique with the lowest production costs was awarded the highest number of points. A lower number of points was then calculated for the other mining techniques, depending on their difference to the most cost-effective technique, which is shown in figure 7. The scaled presentation of the calculations shows vertical ripping as the most cost-effective mining techniques. The distance (or difference) to drilling and blasting amounts to approx. 5 points (i.e. 6 Cent/t) and low in comparison with the other techniques. Cutting with the bucket-wheel excavator takes the third place.

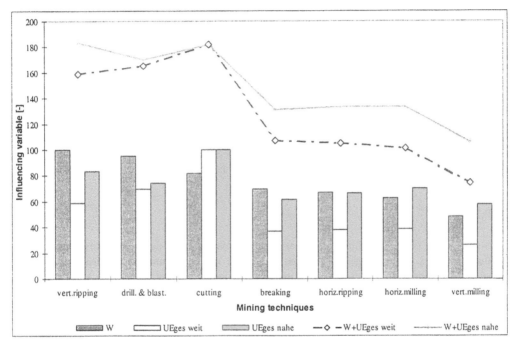

W – economic impact value (production costs); $UE_{ges-weit}$ – weighted environmental effects caused by the far-away open-cast mine; $UE_{ges\ nahe}$ – weighted environmental effects caused the open-cast mine close to sensitive objects ($\approx$300 m); W+UE – sum of the influencing values determined by the economy/commerce and the environment.

Figure 7. Presentation of the environmental effects and the costs after scaling the results

Cutting with a bucket-wheel excavator is the most environmentally-friendly mining technique, when looking at the weighted results of the environmental effects caused in both the far-away open-cast mine and the open-cast mine close to a sensitive object. This can be attributed to the use of electric power in the extraction system. The second place is taken by drilling and blasting in the far-away open-cast mine, and the third place by vertical ripping. If the open-cast mine is closer to a human settlement or to any other sensitive objects, the order of the environmental friendliness changes. While the first place remains unchanged, the second place is taken by horizontal ripping and the third one by drilling and blasting. This is largely due to the vibration caused by the blasting that is clearly higher than that caused by any other mining technique. The high degree of vibration caused by blasting can become a problem the closer the distance between the open-cast mine and the sensitive object concerned is, which, in turn, may jeopardize the continued validity of the extraction license. Although the report of the explosion (the bang) has been disregarded for this assessment, the noise emission values for blasting activities in a residential area, in a mixed area or in a recreational area would have to be observed in any case. As an explosion is a short-term event, the standard emission values may be exceeded in accordance with VDI 2058, sheet 1, by a maximum of 30 dB during the day. A minimum distance of approx. 150 m is required for blasting operations in villages and mixed areas. In accordance with the assessment methodology (as detailed in figure 3), any technique by which a limit is exceeded would then be disregarded.

The comparison of the environmental impact of the individual mining techniques shows that the relative differences of a far-away open-cast mine are much higher than those of an open-cast mine near a built-up area.

Figure 7 makes it clear that cutting, vertical ripping as well as drilling and blasting take a top position among the mining techniques in both cases under examination (a far-away and a close-by open-cast mine) as far as costs and environmental effects are concerned, with drilling and blasting in the vicinity of a town or a village being less favorable than in a far-away place. The most cost-effective mining technique employed in an open-cast mine close to a sensitive object is the second-best as regards its environmental friendliness.

Figure 7 also illustrates a trend indicating a connection between the influential factors determined by the economy and by environmental friendliness. The mining techniques with the lowest environmental impact are also the most cost-effective ones and hence socially most acceptable. This relationship has only been proved in this example and is to be verified by further investigations.

6. SUMMARY

The close vicinity of mining companies to areas given over to other uses, such as residential areas or nature reserves, have triggered a negative attitude of the population towards rock mining companies. Demands for the use of environmentally-friendly mining techniques are increasingly voiced and reference is made to extraction techniques without the use of explosives in this connection.

Based on the definition of what an environmentally-friendly mining technique is supposed to be, this paper presents a methodology that contains technical/technological, economic/commercial and ecological investigations into different mining techniques. When assessing a mining technique in accordance with its environmental friendliness, the analysis should also take social factors into account, apart from economic/commercial and environmental ones. If necessary, these aspects should be differently weighted within the overall assessment and looked at in their entirety. This relationship requires further investigations into and analyses of the rock mining sector.

With the help of the described methodology, six mining techniques in addition to drilling and blasting have been examined with regard to their production costs and the environmental effects caused by SO_2, CO_2, the cumulative energy input (KEA), noise, vibration and dust, in which case a model of a deposit in a fissured rock formation with a pressure resistance of 20 MPa was used. The intention was to gain quantitative data wherever possible. These data were scaled and fed into an evaluation matrix, in which the environmental effects were weighted differently. The results were then assessed in the light of two fundamentally different cases, one being an open-cast mine close to sensible objects, say a town or a village, and the second one being a remote open-cast mine. The local and regional/global environmental effects have been weighted in both cases. As a result of the analyses, the most cost-effective and most environmentally-friendly mining techniques close to and far away from a residential area as well as the most environmentally-friendly mining technique have been established for the two assessment cases by also taking the production costs into account.

Vertical ripping, drilling and blasting as well as cutting proved to be the most cost-effective mining techniques and also the most environmentally-friendly ones in operations with a production capacity of 300 000 t/a.

7. LOOKING AHEAD

Taking a model deposit as an example, this paper introduces a methodology for technical/technological, economic/commercial and ecological investigations. It also discusses the relationship between mining techniques on the one hand and economic/commercial, environmental and social aspects on the other.

It is planned in the future to examine the use of mining techniques in other deposits by taking different properties of the rock formations and production capacities into account. It is intended to verify the relationship between the influence environmental factors have and commercial aspects when rock is extracted in mining operations. The differentiating assessment of the environmental effects and the commercial aspects are intended to be further adapted to each other. In order to carry out an assessment in its entirety, social aspects are to be looked at as well.

Depending on the rock properties in compact rock formations with a high pressure resistance, drilling and blasting seems to be the only reasonable mining technique. In this respect, this methodology offers further ways of examining the extraction system as such more closely in the future.

REFERENCES

GEMIS (Globales Emissionsmodell Integrierter Systeme) Version 4.0. Öko-Institut, Darmstadt 2000, (http://www.oeko.de/service/gemis/)
Niemann-Delius C., Hennig A.: Sprengstofflose Gewinnung − Untersuchung unter besonderer Berücksichtigung ihrer Übertragbarkeit auf die Natursteinindustrie. 1. Teil. In Die Naturstein-Industrie, 40 (2004) 4, S. 24–27.
Schmieder P.: Unterlagen zum Forschungsvorhaben Anwendung and Weiterentwicklung der methodology der Umweltbilanzierung beim Abbau von Festgestein (unveröffentlicht). Institut für Bergbau and Spezialtiefbau, TU Bergakademie Freiberg 2002–2004.
Schmieder P.: Wirtschaftliche Gewinnung einer Gipslagerstätte unter Berücksichtigung von Umwelteinwirkungen. Diplomarbeit. Institut für Bergbau, TU Bergakademie Freiberg 2001.
Schmieder P., Bui Nam: Befahrungsunterlagen über Gewinnungsbetriebe mit sprengstoffloser Gewinnungstechnologie. Unveröffentlicht, TU Bergakademie Freiberg, Institut für Bergbau 2003.
Schmieder P., Drebenstedt C.: Anwendung and Weiterentwicklung der methodology der Umweltbilanzierung für den Abbau von Festgestein. In Sprengstofflose Festgesteinsgewinnung im Tagebau and im Bauwesen. Heft 89 der Schriftenreihe der GDMB. Freiberg 2003, ISBN 3-935797-13-3.
Todzi M.: Mechanische Löseverfahren zur sprengstofflosen Gewinnung von Festgestein unter besonderer Berücksichtigung der schlagenden Gewinnung mittels Großhydraulikhammer. Dissertation. Fakultät für Bergbau, Hüttenwesen and Geowissenschaften der RWTH, Aachen 2000.
Wehrsig H., Drebenstedt C.: Notwendige Untersuchungen für die Bewertung von Umweltauswirkungen beim übertägigen Abbau von Festgestein. In ZKG International, 91 (2002) 3, S.60–67.

International Mining Forum 2005, Sobczyk & Kicki (eds) © 2005 Taylor & Francis Group, London, ISBN 0415 375525

Assessment of the Accuracy of Ground Movement Elements Prediction Using Simulation Method

Wojciech Naworyta
AGH – University of Science and Technology. Cracow, Poland

Joachim Menz
Technische Universität Bergakademie Freiberg. Freiberg, Germany

Anton Sroka
Technische Universität Bergakademie Freiberg. Freiberg, Germany

ABSTRACT: Method to calculate movement elements of ground and ground surface, based on simulation approach was presented in this study. The method assumes that the parameters of the theory describing properties of the ground are treated as random functions. The method allows to calculate the most probable course of any movement element, to assess its confidence interval and probability that the value of movement element will exceed assumed critical values.

1. INTRODUCTION

In any forecast process there is a risk of error of prediction. The value of prediction error depends on various factors such as insufficient understanding of the predicted phenomenon, an error of the model applied to its description as well as an error of parameters used to calculate predicted values. Acceptable range of a prediction error (tolerance) differs between disciplines. In hydro-geological or in atmospheric forecasts tolerance is relatively large. In mine survey, where the influence of underground extraction on the surface needs to be predicted basing on assumed extraction parameters and on assessed parameters characterizing overburden, large prediction error cannot be accepted. The more intensively urbanised is the surface, the more exact the assessment of the exploitation effects on the surface must be.

In the recent decades several methods to predict an effect of exploitation on the surface and on the ground have been developed. The most commonly applied in Poland method by Knothe (Knothe 1953) has a solid theoretical foundation in stochastic medium theory of Litwiniszyn (Litwiniszyn 1956). Basing on the shape of projected excavation void, dynamics of extraction process and knowing the parameters of a calculation method it is possible to predict relatively exactly the subsidence values at specific points on the surface. Comparison of the predicted values with the observed ones indicates some discrepancies. In case of subsidence, the discrepancies are usually negligible, but in case of horizontal movements of the surface the discrepancies may be considerable. An analysis of a large measurement results database indicates that there is a certain natural accuracy limit for the prediction of exploitation effects on the ground and the surface.

The behaviour of ground and its properties are not entirely predictable. Description of a rock mass using a model in which geomechanic properties are described by one or two constant values is an obvious simplification. Consequently, application of such a model will result in distinct differences between the forecasted and the actual reaction of the ground. However, it is difficult to assume that better prediction methods considering all properties of the rock mass will be developed in the future. Ground or its elements react to an excavation void in a way resembling stochastic reaction (Litwiniszyn, 1956), and the methods relying on this assumption (e.g. the method of Knothe) enable to predict the most probable course of this reaction. Therefore, it is difficult to presume that we will be able to predict its reaction more accurately even if we have the exact knowledge of the properties of the ground. For instance, Klein (Klein 1979) reported that despite great care in conducting successive experiments with simple homogenous sand models it was impossible to obtain the same results.

We think that one should not expect that the accuracy of prediction of the exploitation effects on ground surface will be considerably improved in the future. However, the prediction elements may be supplemented with confidence data. Next to the predicted course of particular movement elements of a ground and surface, a confidence interval for prediction and a probability of predicted values to occur should be given. This article presents a possible solution to this problem.

2. STARTING POINT FOR ELABORATION OF A PREDICTION METHOD USING VARIABLE PARAMETERS OF A MODEL

The above-described problem was a starting point to elaborate a method based on assumption that a model parameters have a stochastic character. (Naworyta 2004) during his PhD study supervised by professor J. Menz and A. Sroka worked out algorithms enabling to simulate ground movement elements. The question to be answered by authors is:

How the uncertainty of the assumed parameters characterizing geological and exploitation conditions affects the predicted course of movement elements of ground surface and ground?

Assuming, that the fluctuations (deviations from expected course) of movement elements observed on the surface result only from fluctuations of parameters characterizing geological and mining conditions at the level of an exploited deposit, the above question may be asked in another way:

How large are the fluctuations of parameters characterizing geological and exploitation conditions, which result in fluctuations in the course of the observed surface movement elements?

To answer the above questions a method of simulation of ground reaction to underground exploitation was proposed. The method bases on assumption that the model parameters within the area affected by exploitation do not behave as constant values, but may take values from a certain range. Additionally, it was assumed that these values do not vary within the space under consideration freely, but they are similar to the neighbouring values. For instance the thickness of an exploited deposit after coal extraction may average 2 m and standard deviation may be 10% (±20 cm). Such fluctuations resulting from deposit characteristics and exploitation technique are possible. However, it is unlikely that these values change rapidly (for instance the thickness of deposit at one point is 1,60 m and one meter further changes to 2,40 m). More probable are gradual changes of parameters. Hence, using model based on the theory of Knothe (Knothe 1984) it may be assumed that the thickness of an exploited deposit g, the subsidence factor a, as well as a parameter of ground expressed for instance with the so called radius of scattering of main influences r (or with the angle of the range of influences − β) are regarded as realizations $a(x,y)$, $g(x,y)$ and $r(x,y)$ of Random Functions $A(x,y)$, $G(x,y)$ and $R(x,y)$. The variability of Random Functions is subject to certain laws.

It may be easily justified, that the thickness of an excavation void g is not constant within the exploited deposit. Also the subsidence factor a (actually a function $a(x,y)$) is spatially variable. For instance if the void is backfilled then the density of the fill and its effectiveness will be different at different places. Thus there is a physical explanation of the assumption, that the values of this parameter are spatially variable – similarly to the realization a(x,y) of Random Function A(x,y). The spatially variable values of the ground parameter r(x,y) should not be understood as a merely physical property of the ground related for instance to its geological characteristics. These parameters should be understood also as a random reaction of the ground to the extraction of a deposit element. Thus we can imagine that if it would be possible to repeat the extraction process several times at exactly the same conditions (similarly to the cited study of Klein (Klein 1979)) the reaction of the ground each time would be slightly different, despite that all the time this would be the same ground. Each time the subsidence trough on the surface would have a different radius r_i.

Algorithms based on geostatistical assumptions are suitable for modelling of such complex functions of model parameters. The experiments with digital models (Naworyta, 2003) indicated that if geostatistical dependencies were not considered (meaning that parameters changed stepwise from one point to another without any dependency on their localization) the simulated surface movement elements were not similar to the observed ones.

3. MAIN ASSUMPTIONS OF THE METHOD OF SIMULATION OF SURFACE MOVEMENTS EXERTED BY MINING EXPLOITATION

The principle of the method is the assumption that model parameters behave as spatial variables. Each element of the extracted deposit (referred to as unit deposit element – called further UDE) exerts a unit reaction of the ground, revealed on the surface as an elementary subsidence through. In the mathematical model such a subsidence trough will have a form suitable to the assumed subsidence function (influence function). Elementary subsidence troughs are the functions of spatially variable parameters of the model. Therefore, they differ from each other in terms of volume and radius of scattering of the main influences r_{ij}:

$$s_{ij}(x,y,t) = -a_{ij}g_{ij} \cdot f(x - x_{ij}, y - y_{ij}, r_{ij}) \cdot z(t - t_{ij}) \cdot P_{ij} \qquad (3.1)$$

where $s_{ij}(x,y,t)$ – elementary subsidence trough; a_{ij} – value of subsidence parameter for ij-UDE; g_{ij} – thickness of extracted deposit for ij-UDE; $f(x-x_{ij}, y-y_{ij}, r_{ij})$ – influence function; $z(t-t_{ij})$ – time function; t_{ij} – time of exploitation of ij-UDE; r_{ij} – parameter characterizing ground in exploitation influence area of ij-UDE; P_{ij} – area of ij-UDE and x_{ij}, y_{ij} – coordinates of ij-UDE.

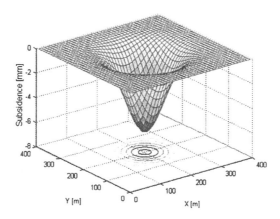

Figure 3.1. Elementary subsidence trough resulting from the extraction of a unit deposit element (UDE)

The model of a subsidence basin $S(x,y,t_{ij})$ is a result of summation of numerous elementary troughs $s_{ij}(x, y, t_{ij})$ (linear superposition of effects). This model is expressed as a matrix:

$$S(x,y,t) = \sum_{i}^{m}\sum_{j}^{n} s_{ij}(x,y,t_{ij}) \tag{3.2}$$

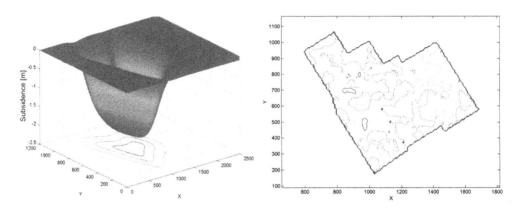

Figure 3.2. Simulated entire subsidence basin (left) and a layout of exploitation longwalls with lines of the simulated maximum possible subsidences s_{max} (right). Visible uneven exploitation boarders result from assuming UDE-area of 10×10 m

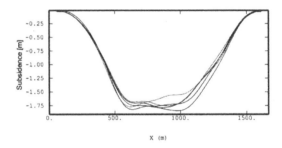

Figure 3.3. Examples of the simulated profiles of a final subsidence trough

The presented example of a simulated subsidence basin (fig. 3.2 left) seemingly differs only slightly from the one simulated using traditional method with constant model parameters (a, g, r – const). However, the enclosed profiles of some simulated subsidence basins (fig. 3.3) show that despite assuming the same simulation parameters the subsidences differ significantly from each other. In order to demonstrate the effects of geostatistically-simulated course of the model parameters on figures 3.4 better, the spatial distributions of residuals were presented (eq. 3.3).

$$R(x,y) = S^{sim}(x,y) - S^{comp.}(x,y) \tag{3.3}$$

where $S^{sim}(x,y)$ – matrix of the simulated subsidence; $S^{comp.}(x,y)$ – comparison matrix of the basin for model parameters given as constant values $a = \bar{a}$, $g = \bar{g}$ and $r = \bar{r}$.

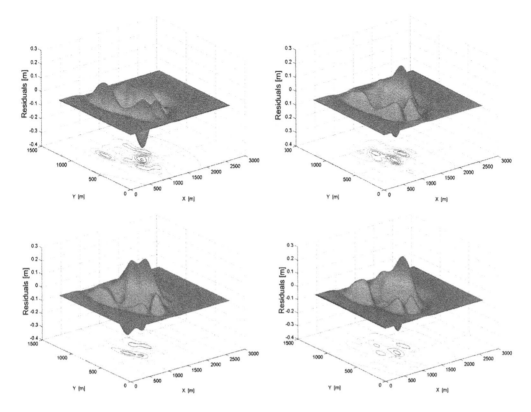

Figure 3.4. Deviations from the course of subsidence of the surface points forecasted
in the standard manner as a result of subsidence simulation using variable model parameters.
The assumed simulation parameters are given in table 3.1 and table 3.2

Table 3.1. Parameters of simulation (fig. 3.4)

	Mean $\overline{x}$	Standard deviation s [% $\overline{x}$]	Model of variogram[1]	Range of variogram [m]
Thickness of the exploited deposit g [m]	2,8	0,14 [5%]	spherical	100
Subsidence factor a	0,7	0,07 [10%]	spherical	100
Angle of the range of influences β [gon]	68	3,4 [5%]	Gauss-model	100

Table 3.2. Other parameters of exploitation and the deposit (fig. 3.4)

Exploitation depth H [m]	Time coefficient c [1/year]	Time t [year]
490	2	10

---

[1] See chapter 4.

127

Interesting is the scatter-plot presenting a dependency of residuals on the depth (fig. 3.5). The graph shows that if ground would behave as a presented digital model, the highest deviations would be expected in the lower part of a subsidence basin. The highest fluctuations of other ground movement elements (resulting from subsidence – a curvature and deformations) should also occur there. In consequence one could conclude that fluctuations in the observed course of deformation within the compression zone should be larger than in the tension zone. This is confirmed on the graph presenting simulated deformations (fig. 3.8).

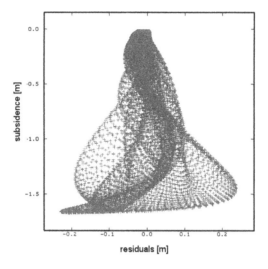

Figure 3.5. Deviations of the simulated course of subsidence from the course forecasted
using the standard method, as a function of subsidence

For the practical presentation of horizontal movement elements of a ground surface, another method was applied. Movement elements will be simulated not in the entire zone of influence of the exploitation but only along a hypothetical measuring line (survey line). Such an approach may only be applied to predict the influence of longwall mining – in the case of coal mining, for instance (fig. 3.6).

In a simplified simulation system, a unit deposit element is assumed as a rectangle (a strip). This method requires assuming that within exploited element of a deposit (e.g. during a shift or twenty four hours) the model parameters are constant. Such a model is inferior to the previous one but this simplification is advantageous because of shorter calculation time and easier interpretation of results. To calculate subsidence along theoretical measuring line the following equation will be applied:

$$S(x,y,t) = \sum_{i=1}^{n(t)} a(x_i) \cdot g(x_i) \cdot \Delta x \cdot f(x - x_i, r(x_i)) \cdot z(t - t_i) \cdot \int_{-\frac{d}{2}}^{+\frac{d}{2}} f(y, r(x_i)) dy \qquad (3.4)$$

where $S(x,y,t)$ – subsidence as a function of position and time; x_i – coordinate of a unit deposit element (UDE) in the form of a strip; $a(x_i)$ – subsidence factor for i-UDE; $g(x_i)$ – thickness of exploitation void of i-UDE; $r(x_i)$ – radius of scatter of the main influences for i-UDE; $f(x-x_i, r(x_i))$ – influence function; $z(t-t_i)$ – time function; d – width of the longwall, and Δx – width of the strip (UDE, see fig. 3.6) (Sroka, 1978).

128

Summation along OY axis is done by using an integral. This requires an assumption that within the exploitation strip the model parameters are constant.

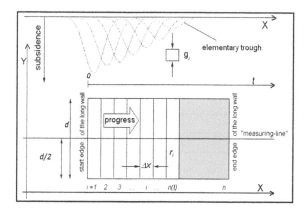

Figure 3.6. Extraction layout and sections of subsidence basins
referring to particular deposit elements (in form of stripes).
g_i – thickness of the deposit within i-UDE; r_i – ground parameter for the influence area of
i-UDE (here: range of scatter of the main influences); d – width of the longwall;
Δx – progress of exploitation within 24 hours (mean width of UDE)

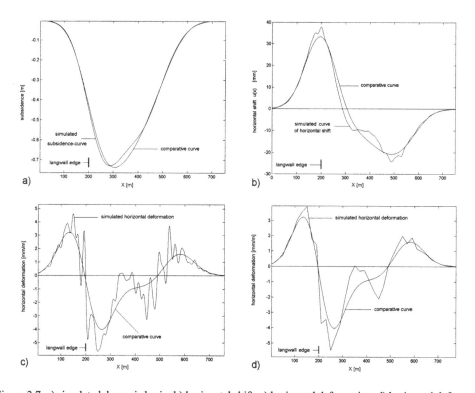

Figure 3.7. a) simulated dynamic basin; b) horizontal shift; c) horizontal deformation; d) horizontal deformation considering the distance between neighbouring measurement points (measuring base l)

129

Figures 3.7a, 3.7b, 3.7c and 3.7d present the subsidence as a dynamic basin and derivatives of the subsidence namely horizontal shift, and relative horizontal deformation (compression and stretching), An exact description of these simulations is given in (Naworyta 2004-2).

To generate graphs presented in figure 3.7 the following parameters were used.

Table 3.3. Parameters of simulation (fig. 3.7)

Parameter	Mean $\overline{X}$	Standard deviation s (% $\overline{X}$)
Subsidence factor a	0,70	0,014 (2%)
Thickness of the exploited deposit g [m]	2	0,04 (2%)
Angle of the range of influences β [gon]	75	1,5 (2%)

Table 3.4. Other parameters of the deposit and the exploitation (fig. 3.7)

Exploitation depth	Measuring base	Width of the longwall	Time	Time coefficient	Number of simulation
H	l	d	t	c	n
[m]	[m]	[m]	[24 hours]	[1/year]	
400	20	150	120	2	100

In the presented examples an influence function according to the theory by Knothe was applied (fig. 3.2, 3.3, 3.4 – eq. 3.5 and fig. 3.7 – eq. 3.6):

$$f(x - x_{ij}, y - y_{ij}, r_{ij}) = \frac{1}{r_{ij}^2} \exp\left(-\pi \frac{(x - x_{ij})^2 + (y - y_{ij})^2}{r_{ij}^2}\right) \tag{3.5}$$

$$f(x - x_i, r(x_i)) = \frac{1}{r(x_i)} \exp\left(-\pi \frac{(x - x_i)^2}{r^2(x_i)}\right) \tag{3.6}$$

and a time function in the form:

$$f(t - t_{ij}) = 1 - \exp(-c_{ij}(t - t_{ij})) \tag{3.7}$$

where c_{ij} – time coefficient characterising the reaction time of the ground to the exploitation of ij-UDE.

To calculate the horizontal movement elements of a ground surface (horizontal shift, horizontal deformation) the relation known from the Knothe theory (Knothe 1984) was applied:

$$u(x) = B \cdot \frac{\partial s(x)}{\partial x} \tag{3.8}$$

where $u(x)$ – horizontal shift and B – proportionality factor.

Parameter B was treated in the simulation similarly to parameters a, g and r as a realization of a random function. The spatial distribution of the function depended on the spatial distribution of random function of r(x):

$$B(x) = 0.4r(x) \tag{3.9}$$

Analyses of dependency (3.8) based on abundant observations indicate, that assumed constant factor B is characterized by a large spatial scatter (Naworyta 2004–2). The simplification was applied (3.9) to present the method more clearly, we think however that in order to achieve more reliable simulation results, a spatial distribution of random function of factor B should be a product of two random functions $\lambda(x)$ and r(x):

$$B(x) = 0.4 \cdot \lambda(x) \cdot r(x) \tag{3.10}$$

where λ – correction factor, which takes the mean value of $\lambda = 0{,}65$ (Sroka 1976).

In the presented simulations it was assumed that a time parameter c is constant within a range of exploitation effects. However, we are aware that this parameter depends on the rock mass properties and similarly to a, g and r should have random values given as a random function c(x,y).

The simulated courses of the movement elements of the surface refer to a single simulation. In order to draw any conclusions from a simulation concerning the probability that certain values of the movement elements of the surface will occur it is necessary to perform a large number of runs with the same initial parameters. Figure 3.8 presents the result of one hundred simulations of the relative horizontal deformation (compression and tension) within an exploited deposit for the final subsidence basin.

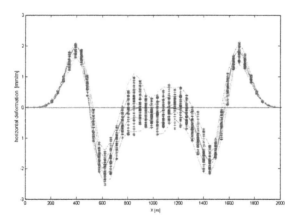

Figure 3.8. Result of one hundred simulations of the horizontal deformation for the final subsidence basin

The graph shows the points referring to simulated values of the horizontal deformation on a theoretical survey base l = 46 m as well as the lines referring to single and double standard deviations of the simulated values. A consequence of the observed differences in occurrence of fluctuations in the subsidence course is that maximal residuals in the course of the horizontal deformation occur also in the lower part of the subsidence basin. In the sections where maximum compression is observed, a large scatter of simulated values occurs as well. Even larger deviations are observed in the middle part of the basin – at places where the value of horizontal deformation should be close to 0.

A practical conclusion from the presented simulation is that the predicted values of horizontal deformation in the compression zone are more biased than the predicted values in the tension zone.

The following parameters were assumed in the presented example (table 3.5 and table 3.6).

Table 3.5. Parameters of simulation presented on figure 3.8

	Mean $\bar{x}$	Standard deviation s [$\% \bar{x}$]	Model of variogram	Range of variogram [m]
Thickness of exploited deposit g [m]	2,8	2%	spherical	100
Subsidence factor a	0,58	2%	spherical	100
Angle of the range of influences β [gon]	62	2%	Gauss-model	100

Table 3.6. Other parameters of the deposit and exploitation (fig. 3.8)

Exploitation depth	Measuring base	Width of the longwall	Time	Time coefficient	Number of simulations
H	l	d	t	c	n
[m]	[m]	[m]	[year]	[1/year]	
500	46	150	10	2	100

Mean values and standard deviations for characteristic points of the graph (fig. 3.8) are shown in table 3.7.

Table 3.7. Mean and standard deviation in the course of the simulated horizontal deformation for characteristic points of figure 3.8

Abscissa [m]	414	644	874	1472	1702
Mean [mm/m]	1,88	-1,97	0,16	-1,73	1,76
Standard deviation [mm/m]	0,12	0,28	0,47	0,23	0,15

4. MODELLING OF PARAMETERS AT THE DEPOSIT LEVEL

Modelling the spatial character of a deposit as well as the spatial course of ground parameters and subsidence parameter a begins with assuming their mean values and variances. Standard deviation can be treated as a measure of assumption uncertainty or a measure of acceptable error of assumption. Basing on the assumed variables a set of supporting points is generated in a coordinate grid. Parameters at the nodes of supporting grid have random values (within the assumed ranges of mean values and their variances) independently of their localization. These points, lying in a grid, will form a base for modelling the parameter courses considering geostatistical assumptions. Points within the supporting grid can be regarded as known values (for instance as values observed in bore-holes). Several methods may be applied to model parameters, including the simplest ones. An example of such a method could be for instance fitting of a trend function to supporting points. Such a function is a polynomial of higher order. Another, more complex method is interpolation (prediction) of a course known as kriging. However, considering the main objective of the presented method it is more reasonable to apply a geostatistical simulation. Both latter methods utilize statistical relationship between the spatial values. This relationship is expressed by a function (a model) of semivariogram. This function (called further a variogram) describes the form and the radius of the statistic correlation between the spatial quantities (fig. 4.1).

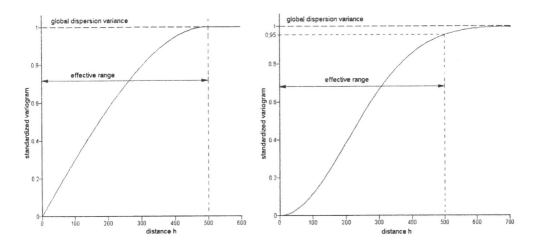

Figure 4.1. Examples of two different standardised variograms spherical model (l) and Gauss-model (r)

Kriging is a method of statistical prediction based on a minimum estimation error as the optimisation criterion. Consequently the most probable course of parameters is obtained, and in geometric expression – the smoothed course. This feature of the presented method should be regarded as a shortcoming. The main objective of the presented method is to show how the modelled course of surface movement elements changes depending on the possible course of function of the geological and exploitation parameters. Thus, not the most probable but the possible course is the point. The possible point is limited by the assumed mean value, the assumed global dispersion variance as well as by the form and range of the variogram. For this reason geostatistical simulation method was applied to model the realization of Random Functions of the geological and exploitation parameters. Contrary to kriging a minimum estimation error of prediction is not assumed.

Figures 4.2 and 4.3 present examples of simulated thickness of a deposit with an application of two different variogram models. The simulation was performed on the support points in the grid of $100 \text{ m} \times 100 \text{ m}$. Assumed mean of thickness $\overline{g} = 1$ m and standard deviation $s_g = 0,1$ m.

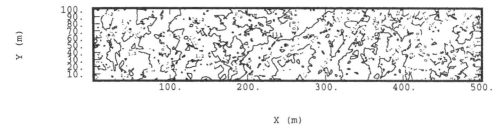

Figure 4.2. Example of simulated thickness of the deposit with a spherical variogram model
with an effective range of 50 m

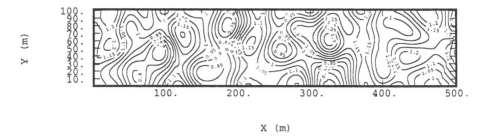

Figure 4.3. Example of simulated thickness of the deposit with a Gauss-variogram model
with an effective range of 50 m

In the presented examples Turning Bands method was applied. This method was developed in the
Centre Geostatistique of Prof. Matheron in Fontainbleau, France (Cressie 1991). A principal program
was written in Matlab language for technical computation. A module of Isatis software (Geovariances,
Fotainebleau) was applied for the simulation (Bleines, Deraisme 2002).

5. DISCUSSION OF THE SIMULATION OF THE MODEL´S PARAMETERS

An analysis of the variograms that where calculated on theoretical mathematical subsidence models
as well as an analysis of the variograms of the observed ground movements` elements that where
made from exhaustive observation material (Naworyta 2004–2) show that there is an easy to forecast,
high correlation between the range of the variogram and the radius of scatter of the main influences r.

However, the results of this analysis do not allow ascertaining the spatial distribution of the mining-
geological parameters at the deposits level. For that, additional experiments based on mathematical
models allowing assuming a certain basic range of the auto-correlation of these parameters would be
necessary. The models presented above assume that if there is a distance of 200 m between support
points, the parameters assumed for these points are not correlated. Therefore, the range of the variogram
that was used for the geostatistical simulation is smaller than 200 m.

An assumed big range of the variogram forces a big distance between non-correlated support points.
Under the assumption of the effective range of the variogram of 1000 m, there would be one support
point in the area of the longwall planned for exploitation and there would be no significant change of the
parameters in the area of the whole longwall. Therefore, the forecast about the exploitation influence on
the ground surface will approximate the classical forecast that is conducted using constant model
parameters (a, g, r = const.).

In the opposite case, where the assumed range of the variogram will be very small (e.g. 20 m; under the assumption that the unit of the deposit is 10 × 10 m), the spatial variation of the parameters will be very high – almost simple random (not correlated) which would create doubts as to the point of conducting geostatistical simulations.

CONCLUSIONS

The intention of the authors was to present a new approach to the problem of methods for forecasting surface deformations caused by underground mining. The main assumption was that the medium (the ground) together with its characteristics imposes constraints on the accuracy of the forecast of its influences on underground exploitation. Therefore, the authors proposed a method that considers imperfect assumptions regarding the models' parameters and that consequently delivers an average course of the surface movement elements together with confidence intervals.

However, it is advised that before the presented method is used, supplementary experiments should be conducted using the digital model and the results compared with the actual, observed ground movement elements. Such experiments would allow adjusting the form and the range of the variograms that have been used in modelling the spatial distribution of the parameters of the model.

REFERENCES

Akin H., Siemens H. 1988: Praktische Geostatistik: Eine Einführung für den Bergbau und die Geowissenschaften. Springer Verlag, Berlin.

Batkiewicz W. 1971: Odchylenia standardowe poeksploatacyjnych deformacji górotworu (poln.). Prace Komisji Górniczo-Geodezyjnej. Geodezja 10. Kraków.

Bleines C., Deraisme J. u. a. 2002: ISATIS, Isatis Software Manual. 4th Edition, Geovariances & Ecole des Mines de Paris, Paris.

Chiles J.P., Delfiner P. 1999: Geostatistics. Wiley Verlag, New York.

Cressie N. 1991: Statistics for Spatial Data. A Wiley-Interscience Publication. John Wiley & Sons, INC., New York u.a.

Deutsch C., Journel A. 1998: GSLIB Geostatistical Software Library and User's Guide. Oxford University. Pr., New York.

Isaaks E., Srivastava R. 1989: Applied Geostatistics, Oxford University. Pr., New York.

Journel A., Huijbregts Ch.J. 1981: Mining Geostatistics. Academic Press, London.

Klein G. 1979: Możliwości określenia stanu deformacji w górotworze naruszonym eksploatacją górniczą, rozpatrywanym jako ośrodek stochastyczny (poln.). Zeszyty Naukowe AGH, Geodezja, Kraków.

Knothe S. 1953: Równanie profili ostatecznie wykształconej niecki osiadania (poln.). Arch. Górn. Hutn., H 1, s. 50–62.

Knothe S. 1984: Prognozowanie wpływów eksploatacji górniczej (poln.). Wyd. Śląsk, Katowice.

Litwiniszyn J. 1956: Gebirgsbewegungen über einem Abbau als stochastischer Prozess aufgefasst. Freiberger Forschungshefte, C 22, s. 45–64.

Litwiniszyn J. 1956: Zastosowanie równań procesów stochastycznych do mechaniki górotworu (poln.). Arch. Górn., H. 3, s. 243–267.

Menz J. 1981: Die Bedeutung der Geostatistik für das Markscheidewesen. Vortrag 14. Fachtagung Markscheidewesen in Gera.

Menz J. 1990: Anwendung der Geostatistik zur Gebirgs- und Lagerstättengeometrisierung. Monographie und Studienhilfe, TU Bergakademie Freiberg.

Naworyta W. 2003: Gebirgskinematische Analyse unter Nutzung der räumlichen Statistik, Abschlussbericht des Graduiertenkollegs Räumliche Statistik. TU Bergakademie Freiberg.

Naworyta W. 2004–1: Geostatistische Genauigkeitsuntersuchungen zur markscheiderischen Vorausberechnung horizontaler Gebirgsbewegungen. 5. Geokinematischer Tag, Schriftenreihe des Institutes für Markscheidewesen und Geodäsie an der Technischen Universität Bergakademie Freiberg. Verlag Glückauf GmbH – Essen.

Naworyta W. 2004–2: Gebirgskinematische Analyse unter Nutzung der räumlichen Statistik. Dissertation der TU Bergakademie Freiberg.

Popiołek E. 1976: Rozproszenie statystyczne odkształceń poziomych terenu w świetle geodezyjnych obserwacji skutków eksploatacji górniczej (poln.). Zeszyty Naukowe AGH, Nr 594, Kraków.

Popiołek E. 1977: Próba oceny dokładności prognozowania maksymalnych poeksploatacyjnych odkształceń poziomych terenu (poln.). Ochr. Teren. Górn, Nr 39.

Sroka A. 1976: Przybliżona metoda określania przemieszczeń punktów górotworu i powierzchni dla małych, regularnych pól eksploatacyjnych (poln.). Zeszyty Naukowe AGH, Geodezja, H. 46, s. 81–95, Kraków.

Sroka A. 1978: Teoria Knothego w ujęciu czasoprzestrzennym. Polska Akademia Nauk, Geodezja, z. 24, Kraków.

Sroka A. 1984: Abschätzung einiger zeitlicher Prozesse im Gebirge. Schriftenreihe Lagerstättenerfassung und -darstellung. Bodenbewegung und Bergschäden, Ingenieurvermessung. Montanuniversität Leoben, Leoben.

Sroka A. 1999: Dynamika eksploatacji górniczej z punktu widzenia szkód górniczych. Studia, Rozprawy, Monografie, Nr 58, Polska Akademia Nauk, Instytut Gospodarki Surowcami Mineralnymi i Energią, Kraków.

International Mining Forum 2005, Sobczyk & Kicki (eds) © 2005 Taylor & Francis Group, London, ISBN 0415 375525

Simple and Advanced Methods of Mineral Projects Risk Analysis – Application Examples

P. Saługa
Polish Academy of Sciences, Mineral & Energy Economy Research Institute. Cracow, Poland

E.J. Sobczyk
Polish Academy of Sciences, Mineral & Energy Economy Research Institute. Cracow, Poland

ABSTRACT: Uncertainty and risk are inherent to any mining project. Identification of sources of uncertainties and assessment of risk levels are some of the most important issues that contemporary companies have to deal with. The problem specifically concerns the mining industry where the number of uncertainties and the risks associated with them is exceptionally high. That is why scientific research to develop methods to assess and quantify risk as accurately as possible has been conducted for several years. The paper presents some of the popular (risk-adjusted discount rate, sensitivity analysis, scenario analysis) and advanced (Monte Carlo simulation, decision-tree analysis) project risk analysis methods.

INTRODUCTION

No mining company should start with any substantial mining project without prior conducting a risk analysis to help in the decision making process.

Risk analysis means any method – quantitative and qualitative – allowing assessing levels of risk associated with specific situations.

The aim of a risk assessment is not to reduce risk – this is not to be accomplished by a mere analysis – but rather to enhance the managers' knowledge and understanding of risk in order to take appropriate actions (Torries 1998).

According to Simonsen and Perry (1999) risk analysis is a process of identifying potential consequences of decisions made and its most important goal is to predict the probability of achieving the planned profit in a given period of time. The answers to be found by a risk analysis are thus as follows (Simonsen, Perry 1999):

1) What can happen?
2) What is the probability of it happening? and
3) What would the consequences of such an event be?

Risk analysis may be regarded as an analytical and opinion making task, whose goal is to assess risk inherent to key variables of a project (Helfert 2000). Such an analysis may take various forms.

According to Runge (1998), a risk analysis encompasses the use of certain techniques allowing assessing risk on the basis of the assumed distributions of cost and income uncertainties. The process involves random and simultaneous changing of several input parameters; it is done, however, with regard to the probability of their occurrence.

From the point of view of the involvement risk has in the process of making decisions, risk assessment methods can be divided as follows (Kasiewicz, Rogowski 2004):

1) Direct methods, where risk is directly included in the decision criterion for an economic viability assessment method, which makes them become one of the criteria of the decision making process (for example risk-adjusted discount rate, payback period, certainty equivalent).

2) Indirect methods serving to obtain additional information about levels of risk, and so not included in the decision criterion itself; these methods do not directly become criteria of the decision making process (sensitivity analysis, scenario analysis, statistical analysis).

Among the methods named above, the following are regarded as standard in the assessment of discounted cash flow risk:

1) Risk-adjusted discount rate method.
2) Sensitivity analysis.
3) Scenario analysis.

Advanced risk analysis methods include:

1) Monte Carlo simulation.
2) Decision tree or Bayesian analysis.

RISK-ADJUSTED DISCOUNT RATE METHOD

Two basic indicators influence a businessman's decision with regard to an investment:

1) Bank interest rates, which are possible to obtain (it is obvious that only if the rate of return on the invested capital is higher than the bank interest rate a businessman will be interested in investing his money in the project).

2) Projected profit rates for various alternatives of the project.

Discount rate defines the relationship between the present and the future value of money. Runge (1998) defines it as an interest rate that allows a company to assess values of future events in comparison to the current circumstances. Lawrence (2000) says that discount rate represents a capital return rate required by investors to assess investment needs of projects, and dependent on the cost of the capital and the risk of that project. Torries (1998) on the other hand regards it as an interest rate used to re-calculate (discount) value of future profits and costs in terms of current money.

Taking the above into account discount rate may be defined as a yearly payment for an accumulated consumption, expressed as an interest on a borrowed capital (equal to the accumulated consumption), which compensates the lenders for making the necessary investment capital available (Jankowska-Kłapkowska 1992), (Pazio 2002).

A discount rate assumed for a project encompasses three basic elements (fig. 1) (Smith 1994, 2000):

1) "risk-free" interest rate (approx. 2.5% for the USA and Canada);
2) project-specific risk (3–16%);
3) country risk (0–14%, country-dependent).

In current money it additionally includes:

4) inflation rate.

In practice, investors express their opinion of the risk of a specific project by adopting a specific discount rate, so the point of the risk-adjusted discount rate method lies in choosing a rate appropriate to the perceived level of risk. Choosing the correct rate is a complicated process. It is done by an experienced analyst who increases the discount rate by a value closely related to the risk, which is subjectively perceived by him. The value may also be mathematically calculated with the use of a model (i.e. – capital asset pricing model CAPM).

In practice most companies adjust to risk by using a so-called hurdle rate, which regulates the top limit of discount rate (Moyen et al. 1996). Even though the limit changes from one company and project to another, the most common value used for feasibility studies in the North-American (the USA and Canada) base metal industry is 10% (for prices expressed in constant dollars) (Smith 1994). A poll done among the members of the Canadian Institute of Mining, Metallurgy and Petroleum

(CIM) – Mineral Economics Society by Smith caused this value to be increased by 1.3% (Smith 2000). Other authors use similar values – Davis (1998) assumes 8–12%, and Moyen et al. (1996): 5–10%.

In view of the above, the used discount rate decreases as a project progresses into higher design stages. For base metal mining Smith (1994, 2000) gives the following values:
- at the stage of early exploration – approximately 17.5%;
- during pre-feasibility study – approx. 13.5%;
- during feasibility study – approx. 11.5%;
- during operation – approx. 8.5%.

It has to be stressed at this point that the quoted discount rate of 10–11.5% used by mining companies to assess risk during feasibility study in countries characterized by low political risk (the USA, Canada) is calculated under specific assumptions, as follows (Smith 1994, 2000):
1) Zero inflation rate (constant dollars).
2) Financing the project with 100% of equity capital.
3) Calculations done after tax.

The rate of 10–11.5% is used by most mining investors to make definite decisions; as millions of dollars are at stake, this value seems to be generally regarded as valid.

For the sake of simplicity, the discussion on component elements of discount rates so far in the paper was limited to projects sited in countries of low political risk: the USA and Canada. When investing in countries of higher political risk one should increase the discount rate by a certain value depending on the position such a country takes in the ratings done by specialized international agencies and banks. Such premium may range between values close to zero and 10%, even though it may be as high as 13–14% in some cases (Smith 1994, 2000). In each case it is necessary to assess the influence country-specific risk parameters may have on the project and the discount rate (and so the project's valuation). The result of such assessment is reflected in the discount rate.

Risk-adjusted discount rate method has the following advantages:
1) It is a simple and direct reflection of the perceived risk of the project.
2) It's useful in the project valuation process.
3) It's useful in comparing project alternatives.

Its disadvantages can be named as follows:
1) Difficulties in establishing a correct discount rate due to:
   a) an element of subjectivity in selecting discount rate at an *ad hoc* basis;
   b) shortcomings of analytical models.
2) Lack of information regarding distribution of investment risk.

SENSITIVITY ANALYSIS

Sensitivity analysis is a simple method of risk assessment consisting in examining the influences of changes (that may take place in future) of key variables considered in the project effectiveness analysis on the project's viability (or rather: on one of its effectiveness indicators – i.e. net present value NPV or internal rate of return IRR). According to (Kasiewicz, Rogowski 2004) sensitivity analysis is the most popular method of analysis of investment risk.

In its simplest form sensitivity analysis examines the influence percentage deviation of an argument (describing variable) has on the result (described variable). The examination consists in assuming certain percentage range of deviation (i.e. from –30 to +30%) of the describing variables from their base (projected) values and assessing the project's viability at these values. Subsequent calculations are done by changing the base level of a selected describing variable in pre-determined steps (i.e. every 1, 5 or 10%). As mentioned before, the values of only one of the independent variables are changed (the values of the other remain at their base levels). It must be noted that with changes of independent variables, dependent variables may change automatically.

A simple tool of sensitivity analysis allowing observing the influence of the examined parameters on a project's effectiveness is sensitivity diagram called also a spider diagram. The diagram may be used to examine which of the parameters have the biggest influence on the changes of the parameters that describe financial effectiveness – NPV, IRR, discounted average cost, time of return on capital etc. Parameter changes are marked off at the 0X axis of the diagram (i.e. by increasing or decreasing their base values by 10, 20, 30% etc) and changes of effectiveness indicator (IRR or NPV) at the 0Y axis.

The influence of each parameter is measured by the angle between the graph line and the 0X axis. This is a so-called sensitivity ratio.

Sensitivity analysis may also be conducted in a manner allowing answering the following questions:

1) What level of the examined variable still ensures project's viability?

or

2) What are the allowed variances of individual parameters still ensuring project's viability?

Risk levels obtained from a risk analysis conducted to help in the process of making decisions with regards to an investment project are presented to the management in the following format (Kasiewicz, Rogowski 2004):

- Sensitivity of the method's absolute decision criterion to changes of the parameters considered in the method.
- Boundary values.
- Safety margins.

Presented below is an example of use of sensitivity analysis. The example relates to a coal-mining project. The project involved longwall exploitation of a mining field with reserves of 4.9 million tonnes. The assumed extraction rate was accepted as 36.120 thousand tonnes per month. It was also assumed that the coal price obtained by the mine was 240.30 Polish zlotys per tonne and its operational costs were 142.30 zlotys per tonne (1 Polish zloty ≈ 1/3 US$). The required investment capital was estimated at 13.490 million zlotys.

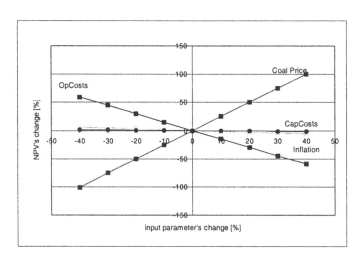

Figure 1. Sensitivity analysis (spider diagram) of a mining project

The sensitivity analysis involved examining the influence of the abovementioned variables, whose values were increased or decreased by 10, 20, 30 and 40% (with values of all other variables remaining at constant levels) on NPV. NPV calculation was done for each option.

It is easy to see what levels of risk are attached to each variable simply by looking at the results (fig. 1). As can be seen the project is most sensitive to the coal price and operational costs: price increase by 40% results in a 100% increase of the NPV (and the other way around: reduction of price by 40% decreases the NPV by nearly the same percent); cost increase by 40% results in a drop of the NPV by nearly 60% (assuming other variables remain constant).

As a surprise came a relatively small sensitivity of the NPV to the changes in capital expenditure and the volume of output.

Advantages and disadvantages of sensitivity analysis are presented in table 1.

Table1. Advantages and disadvantages of sensitivity analysis (Kasiewicz, Rogowski 2004)

ADVANTAGES	DISADVANTAGES

1) Is used to analyse risk, indicating areas that should be subjected to more advanced analyses.
2) Is useful in examining new types of projects, whose risks have not been analysed before and where there is no past experience of similar undertakings.
3) Its results may be used in other risk analysis methods.
4) Facilitates direct comparison of risk determined by different independent describing parameters (due to the possibility of presenting all obtained sensitivity graphs on one diagram).
5) Provides – thanks to sensitivity graphs – useful information about the boundary conditions at which decision criteria change, and allows calculating margins of safety.
	1) Assumes the existence of a simplified and frequently untrue ceteris paribus condition (that only one, the examined, independent describing parameter changes and the other remain constant).
	2) Is not complete because investment risk also depends on:
	  – sensitivity of the decision criterion to changes of independent describing parameters, and
	  – range of probable values of these variables illustrated by their probability distribution and only the first factor is taken into account in sensitivity analysis.

SCENARIO ANALYSIS

Scenario analysis is yet another commonly used tool allowing analysing investment risk by deterministic calculus of discounted cash flows analysis (Torries 1998), (Runge 1998), (Wanielista et al. 2002) and others. Contrary to sensitivity analysis the method facilitates changing multiple base parameters within one scenario. Scenario constitutes a projection of the future within the range of the variables describing the econometric model in which a mining project feasibility rating is a described variable. It must be stressed that a "scenario" is not equivalent to a "prediction". It rather is a tool allowing understanding the risk better – a manager using it can say, "I am prepared for any circumstances" (Runge 1998). Scenario analysis was developed after World War II as a method of warfare planning; it took a new shape, however, in the 70's and 80's of the last century.

Three scenarios, i.e. possibilities of certain values of project's parameters to occur are normally defined:

1) Best-case scenario, constructed basing on the most optimistic parameter value estimates; this case in a sense reveals the biggest amount of information about the project.
2) Base-case scenario, constructed basing on the estimates as to the expected values of parameters.
3) Worst-case scenario, constructed basing on the most pessimistic parameter value estimates.

Summa summarum – best-case scenario occurs when all uncertain variables have the best possible values; worst-case scenario – when their values are least favourable.

Base-case (or "expected") scenario occurs when the uncertain variables of the project assume values predicted ("expected") by the analysts and engineers; base case is constructed on the basis of "best" estimates of a project parameters that is why the obtained NPV is often – mistakenly – regarded as the project's expected NPV (statistically, base case is probably not the expected option; the expected case may only be found by conducting a probabilistic analysis, which is not done here). That is why the real expected value should be estimated by some other method than the one presented here.

An example of practical use of scenario analysis presented in the paper is based on the same case study as the one of the sensitivity analysis above. The options shown in table 2, base, best and worst, were constructed on the basis of mining information and analyses.

Table 2. Scenarios of a mining project assessment model

PARAMETER	WORST CASE SCENARIO	BASE CASE SCENARIO	BEST CASE SCENARIO
Coal price [zlotys per tonne]	237.90	240.30	240.90
Operation cost [zlotys per tonne]	267.40	142.30	98.60
Monthly output [tonnes per month]	18,900	36,120	53,550
Investment capital [zlotys]	14,125,000	13,419,000	12,713,000
Inflation rate [%]	8.00	3.00	1.5
NPV [zlotys]	(45,201,598.86)	101,754,408.00	263,985,634.55

Results of the scenario analysis are presented in figure 3. The net present value of the project in base ("expected") case was positive at over 101 million zlotys. In best case the NPV is much more positive – with value about 264 million zlotys. In worst case the result is negative with the NPV about 45 million zlotys below null. The value of the IRR for best case is 22.56%, which is very promising.

Table 3 shows advantages and disadvantages of scenario analysis.

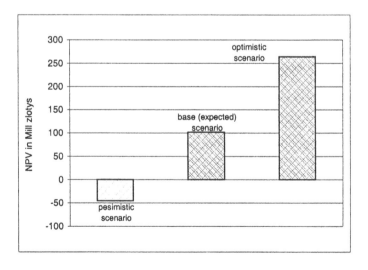

Figure 3. Results of a scenario analysis of a mining project

Table 3. Advantages and disadvantages of scenario analysis (Mielcarz 2004), (Torries 1998)

ADVANTAGES	DISADVANTAGES
1) Relative ease and simplicity of calculations. 2) Possibility of including changes of many base parameters in one scenario. 3) Possibility of establishing logical interdependencies between various parameters. 4) Possibility to construct procedures representing given situations including actions to be taken by managers under the circumstances. 5) It is a semi-estimate of realistic conditions.	1) Allows to allocate a probability to the whole scenario, but the result does not include values of probability for individual variables or events. 2) Decisions made are supported by a limited number of scenarios – big chances of making a mistake.

MONTE CARLO SIMULATION

In the case of a standard deterministic scenario analysis the risk associated with uncertainty of input parameters is normally represented by three main scenarios of future events base (most probable), worst (pessimistic) and best (optimistic).

Scenario analysis is commonly used due to its ease and simplicity of calculations. A question of the number of scenarios needed to be constructed became relevant when researching the method. There were, obviously, views expressed that the best way to assess risk would be to examine as many probable scenarios as possible, or even all of them. Such possibility is afforded by a probabilistic analysis with the use of a method called "Monte Carlo" which was devised by a distinguished Polish mathematician Stanisław Ulam. The name itself, "Monte Carlo", given by Ulam and von Neumann, was used as a cryptonym during the research on atomic bomb (the Manhattan Project). The principle of the method (based on the idea of the Markov's principle) lies in an assumption that many of the relations occurring in circumstances not lending themselves to probabilistic descriptions are easier to assess by stochastic experiments than standard analytical methods. The Monte Carlo method may be in fact regarded as a perfect scenario analysis method because it examines nearly all possible scenarios simultaneously. Uncertain parameters (variables) of the project are input into the method's calculus as probability distributions. One can say that the determined in advance values of parameters are replaced by objective information in a form of their probability, that is a range of possible values of the parameters together with the probabilities of them occurring. Such an approach allows obtaining objective information representing the whole range of possible results of a project and assisting in making decisions in a risk environment.

A detailed description of the use of a stochastic simulation for assessment of viability of a mining project has for the first time been presented by Cavender (1992).

The method consists of the following steps:
1. Constructing of a spreadsheet model.
2. Identifying the most important input parameters and defining the output parameters.
3. Preparation of the data for simulation.
   3.1. Defining probability distribution of each and every essential input parameter and matching them to distinct statistical distributions.
   3.2. Identifying dependent variables and relations between them.
4. Simulation.
   4.1. Generating random values of input parameters from their probability distributions.
   4.2. Successive inputting of the generated random values into the cash flow spreadsheet, calculating outputs and entering them into a results table.

5. Analysis of the results.
   5.1. Interpreting the obtained statistical distributions (and standard deviations) of individual output parameters.
   5.2. Analysis of probability distributions of their possible values.
   5.3. Sensitivity analysis.

In a situation where most of the uncertain parameters can be expressed as probabilities of their possible values the value of discount rate used for a Monte Carlo simulation should be close to the "risk-free" rate. It must be stressed, however, that the expected value of NPV distribution obtained as a result of the analysis is not equal to the price the investor would currently pay for the project. This value cannot be obtained directly from a probabilistic simulation without taking into account the investor's willingness to take risk. Investor willing to invest in a particular project would include risk into the probabilistic analysis in order to assess the project's certainty equivalence which represents the price the investor would be willing to pay (Walls, Eggert 1996), (Torries 1998). Probability analysis describes risk in a simple manner so the investors preferences with regard to risk help to interpret it and determine the price the investor would pay.

An example of using Monte Carlo simulation to model economic viability of a coal-mining project is presented below.

Coal mine X considers a project with some level of uncertainty. The management inquires about the probability of achieving a positive NPV. All the uncertain parameters used in the calculation (and their probability distributions) are listed in table 4.

Discount rate value was accepted as 8.0% due to the fact that most of the input parameters were expressed as probability distributions. Inflation was not taken into account in the calculations.

Table 4. Probability distributions of input parameters for Monte Carlo analysis

VARIABLE	DISTRIBUTION	EXPECTED VALUE
Coal price [tonnes]	Triangular (180.2; 240.3; 300.4)	240.300
Operating cost [zlotys per tonne]	Triangular (98.6; 142.3; 267.4)	169.430
Capital costs [%]	Triangular (3,182,985; 3,536,650; 3,890,315)	3,536,650
Longwall monthly advance [meters]	Triangular (28,896; 36,120; 43,344)	36,120
Reserves per longwall [tonnes]	Triangular (4,263,000; 4,900,000; 5,537,000)	4,900,000
Deposit width [meters]	Triangular (80; 113; 140)	113,000
Inflation [zlotys per meter]	Exponential (3.0)	3.0
Extraction loss ratio [%]	Triangular (16.0; 20.0; 24.0)	20.0
Tailings storage costs [zlotys]	Triangular (12.3; 15.3; 18.4)	15.3
Process recovery [zlotys]	Triangular (94.09; 97.00; 99.91)	97.0

4,000 spreadsheet calculations were done during the simulation. Resultant cumulative probability distribution of the NPV is shown in figure 4. The mean value of NPV distribution is positive (94.302 million zlotys). Figure 4 shows that with the assumed values of input parameters there is a 92.8% chance that the project will be economically viable. The obtained result allows the management to make proper decisions, which will still depend on their acceptance of risk.

Advantages and disadvantages of Monte Carlo simulation are listed in table 5.

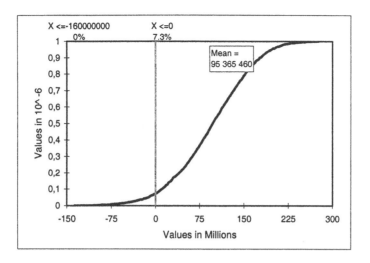

Figure 4. Cumulative probability distribution of NPV

Table 5. Advantages and disadvantages of Monte Carlo simulation

ADVANTAGES	DISADVANTAGES
1) Complete statistical information on project economics. 2) Good estimation of the calculated effectiveness indicator and supplying statistics of its distribution. 3) Understanding of the range of uncertainty associated with the expected value of each indicator. 4) Possibility to estimate probability of certain values of the calculated indicator to occur. 5) Possibility to use "risk-free" discount rate. 6) Possibility to conduct scenario and sensitivity analyses.	1) Problems with constructing probability distributions of uncertain input parameters. 2) Difficulties with determining correlations between input parameters. 3) Problems with estimating project's value (necessity to use "risk-free" discount rate or the preference theory). 4) Difficulties with interpreting the results.

DECISION TREES

Decision tree method is one of the most modern methods used for analyses of investment project risk. It consists in examining sequential decision making processes i.e. such where decisions are made in a sequence and so the next one depends on the one made before. Decision trees help to structure complicated problems and their results' graphic representations resemble trees, hence the name (Szapiro 2000).

A decision tree consists of junctions and branches. Junctions are points in time when events take place. An event may constitute a decision selected from several alternatives represented by a decision junction. Such junctions are graphically shown as rectangulars. Oval junctions express a state of affairs, which does not lend itself to full control of the manager. In this case a probability of an event-taking place is estimated. Such junctions are called random junctions. Branches join all junctions. A situation where there no longer exists a possibility of further action is called an end junction and is shown as a triangle (fig. 5).

Decision tree allows addressing complicated problems. It helps to organise data, makes it easier to have an overlook of the problem, defines optionality of actions and expected reactions at a certain probability of them occurring and attaches costs to every decision made.

The necessity to use substantial quantities of data and probabilities coupled with complicated calculations are the reason why accurate analyses of problems are frequently aborted for the sake of intuitive solutions. Problems as these may be approached with the help of charts called decision trees. As a result of defining the problem and constructing a tree representing it a simple tool helping in the process of making decisions is obtained. The tree makes it easy to determine sequential decisions and examine their sensitivity to parameter changes. Such situation is spoken of as decision-making in conditions of uncertainty. The uncertainties are related to the project's environment.

Classic examples in which the potential of decision trees demonstrates itself are mining projects such as prospecting, exploration, development and extraction of deposits. An example of assessment of such a project, encompassing development of a new field in a coalmine is presented below.

A deposit has been explored to the extent allowing planning 4 longwalls. Total reserves that can be extracted under such design amount to 3.2 million tonnes. Conducting a better geological exploration may answer the question whether two additional longwalls can be sited in the area. This will also allow decreasing investment risk, which is high due to high capital expenditure. The cost of additional exploration by longholes or drives has been estimated at 15 thousand EUR (67.5 thousand zlotys).

Opening a fifth longwall would increase the available reserves to 4.3 million tonnes and with six longwalls it would be possible to mine 5.4 million tonnes. All the longwalls are similar: maximum width – 1.7 meters, length – 250 meters, run – 1,500 meters.

The investment's economic viability was assessed by means of its NPV. The following general assumptions were made (Sobczyk 2002):

- Development will take 3.5 years and will be done at a cost of 4.6 million EUR (20.7 million zlotys).
- Income is calculated under the assumption that the whole extracted coal is sold.
- The assumed operational cost is 32.00 EUR per tonne (144.00 zlotys per tonne) and the coal price is 34.80 EUR per tonne (156.60 zlotys per tonnes).
- Discount rate is 10%.

The first of the longwalls will start production in the middle of the forth year of the project with the others following in succession. It will be possible to start two additional longwalls after positive results have been obtained from geological exploration.

Financial assessment of the project was done for two options. NPV was calculated for both.

The first option assumes that production in the new field will be done at the mine's current level of technical and logistical potential. Effective time of work of a shearer is 3 hours per shift in a 3-shifts-per-24h system and its working speed is 4.2 m/min. From these follows daily production of 2700 tonnes and monthly output of around 60 thousand tonnes. The assumption means that the mine does not invest in face equipment at all, and only the partially amortized machinery already in the mine's possession will be utilized. At the assumed extraction rate the time of operation and the NPV are respectively:

NUMBER OF LONGWALLS	TIME IN OPERATION	NPV
4	5 years and 8 months	EUR 240,000 (1.080 million zlotys)
5	7 years and 3 months	EUR 380,000 (1.710 million zlotys)
6	8 years and 8 months	EUR 720,000 (3.240 million zlotys)

Second option foresees increasing concentration of production. The need for a more efficient shearer use achieved by increasing its effective working time to four hours per shift is stipulated. Organisational changes and introducing an additional production shift are envisaged. The mine invests 3 million EUR (13.5 million zlotys) in a new longwall shearer and an armoured face conveyor. Because of this the shearer's working speed was taken to be 5.5 m/min. Under these assumptions the mine will be able to achieve daily production from one longwall of around 6200 tonnes, which amounts to 130.4 thousand tonnes per month. At the assumed extraction rate the time of operation, comprising equipping, extraction and closure of the faces, and the NPV are respectively:

NUMBER OF LONGWALLS	TIME IN OPERATION	NPV
4	2 years and 8 months	EUR 370,000 (1.665 million zlotys)
5	3 years and 6 months	EUR 410,000 (1.845 million zlotys)
6	4 years and 3 months	EUR 1,200,000 (5.400 million zlotys)

The multitude of options available in the case presented above makes it necessary to conduct a decision analysis. Such an analysis will allow finding an optimal path on the decision tree and calculating possible results. Choosing the best decision consists in applying the maximum expected value criterion.

The following assumptions originating from the decision making model described above were made for the construction of the decision tree:

1. Two options for equipping the faces are considered:
– Equipment partially amortized.
– State-of-the-art longwall shearer and face conveyor.
2. Three design options possible depending on the mining field size:
– 4 longwalls.
– 5 longwalls.
– 6 longwalls.
3. Capital investment cost is very high. Additional exploration needs to be done in order to decrease risk. Results of the exploration may be as follows:
– Plan 4 longwalls – reserves 3.2 million tonnes.
– Plan 5 longwalls – reserves 4.3 million tonnes.
– Plan 6 longwalls – reserves 5.4 million tonnes.

Table 6. Calculating probability of random events

POSSIBLE STATUS	ESTIMATED RISK LEVEL	CONDITIONAL PROBABILITY OF OBTAINING THE RESULT	TOTAL PROBABILITY	BAYESIAN PROBABILITY
Probabilities for 4 longwalls – reserves 3.2 million tonnes				
4 longwalls	0.5	0.6	0.3	0.698
5 longwalls	0.3	0.3	0.09	0.209
6 longwalls	0.2	0.2	0.04	0.093
Total			0.43	
Probabilities for 5 longwalls – reserves 4.3 million tonnes				
4 longwalls	0.5	0.3	0.15	0.417
5 longwalls	0.3	0.5	0.15	0.417
6 longwalls	0.2	0.3	0.06	0.167
Total			0.36	
Probabilities for 6 longwalls – reserves 5.4 million tonnes				
4 longwalls	0.5	0.1	0.05	0.238
5 longwalls	0.3	0.2	0.06	0.286
6 longwalls	0.2	0.5	0.1	0.476
Total			0.21	

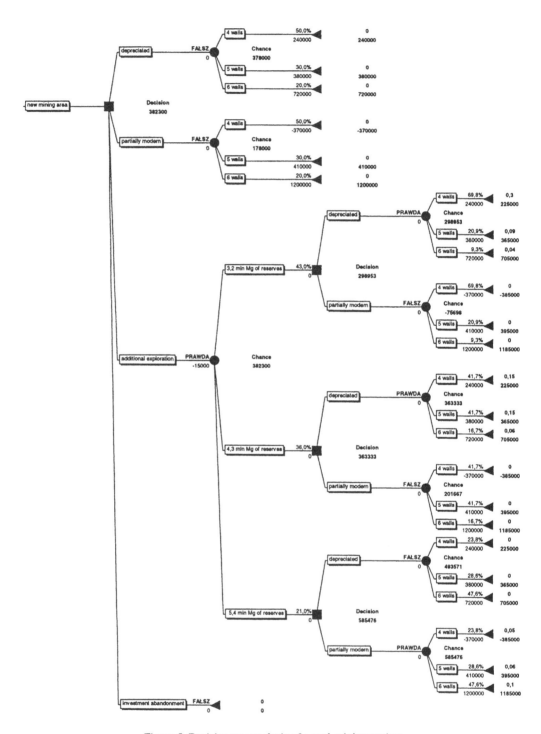

Figure 5. Decision tree analysis of a coal-mining project

Accuracy of assessing the probability is essential when constructing a decision tree. It is specifically important when calculating the expected value. The manner in which values of probability are calculated at all random junctions is shown in table 6.

Interpretation of the tree for example II should lead the investor to ordering additional exploration and following its results:

a) If the exploration reveals a possibility of sitting four longwalls, i.e. mining reserves of only 3.2 million tonnes – the longwalls should be equipped with partially amortized machinery from other sections of the mine.

b) If it is possible to mine 4.3 million tonnes of coal with 5 longwalls – the longwalls should also be equipped with partially amortized machinery from other mining sections.

c) If the exploration reveals a possibility to mine 5.4 million tonnes of reserves with 6 longwalls – the longwalls should be equipped with modern shearer and face conveyors and auxiliary machinery to be reclaimed from other sections.

The above example is presented in the form of a decision tree in figure 5.

The decision tree embodies complete solution to the examined problem. As can be seen, there are 24 possible final situations. Each branch at every random junction is characterised by a probability of its occurrence determined by one of the methods described in example I.

Expected values are then calculated at successive random junctions for every financial result and probability at each branch of the tree. Following the path of maximum expected value arrives at the best decision.

Statistics report (table 7) is a completion to the graphic form of decision trees.

Table 7. Statistics report for the evaluation of an investment project concerning development of a mining area in a hard coal mine

PrecisionTree Statistics Report
For nowe pole eksploatacyjne of kopalnia wegla kamiennego.xls
Created on 2002-07-01 at 11:53:22

Decision	1 : depreciated		2 : partially modern		3 : additional exploration		4 : investment abandonment	
STATISTICS								
Mean	378000		178000		382300		0	
Minimum	240000		-370000		-385000		0	
Maximum	720000		1200000		1185000		0	
Mode	240000		-370000		225000		0	
Std Dev	181427,7		612532,4		339852,5		0	
Skewness	1,119633		0,587223		0,783144		0	
Kurtosis	2,692711		1,876436		4,513206		0	
PROFILE:								
#	X	P	X	P	X	P	X	P
1	240000	0,5	-370000	0,5	-385000	0,05	0	1
2	380000	0,3	410000	0,3	225000	0,45		
3	720000	0,2	1200000	0,2	365000	0,24		
4					395000	0,06		
5					705000	0,1		
6					1185000	0,1		

The statistics report shows that, taking into consideration the EMV value only, additional exploration of the deposit before beginning its mining is the optimum decision.

Figure 6 shows a risk profile for the example under consideration. The profile shows probabilities of obtaining specific results depending on decisions made. I can be seen that the risk of suffering a loss to the amount of 390 thousand EUR (1.8 million zlotys) in the case of deciding to invest in new face equipment without any additional exploration done beforehand is very high at 50%. On the other hand, deciding to go ahead with the project after exploration works have improved the knowledge of the reserves results in a 45% chance of achieving profit of approx. 225 thousand EUR (1 million zlotys).

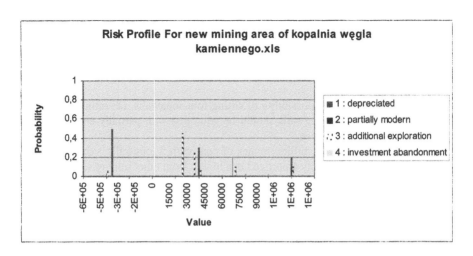

Figure 6. Risk profile for a coal-mining project

A more in-depth risk analysis for the already chosen best decision can be done by interpreting the cumulative distribution of probability of achieving the result, as shown in figure 7. It can be seen that additional exploration ensures favourable end result and decreases the risk level. The graph shows that there is a chance of over 80% that the final financial result will reach 400 thousand EUR (1.8 million zlotys).

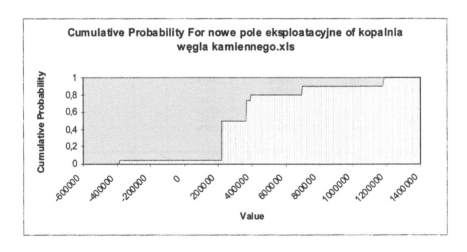

Figure 7. Cumulative probability distribution for a coal-mining project

Figure 8 shows the graphic result of a sensitivity analysis. The diagram shows which of the values has the largest influence on the result of our decision. In our case, the sensitivity analysis describes the sensitivity of EMV to changes of individual factors influencing this quantity. It shows the most critical factors influencing the evaluated quantity, but does not inform about the range of their acceptable changes.

150

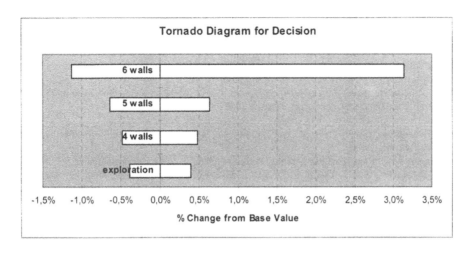

Figure 8. Sensitivity analysis (TORNADO type) for the selected decision of the investment project concerning development of a mining area in a hard coal mine

In case of TORNADO type diagrams, the longer column both in the direction of negative and positive values, the more sensitive to changes of this variable the investigated quantity is. In our analysis, a change of NPV for the project with 6 walls in the mining area exerts the largest influence on the value of EMV. Even a small change of NPV for this project leads to significant changes of EMV.

REFERENCES

Cavender B.W.: Determination of the Optimum Lifetime of a Mining Project Using Discounted Cash Flow and Option Pricing Techniques. Society for Mining, Metallurgy, and Exploration, Inc., Preprint No. for presentation at the SME Annual Meeting, Phoenix, Arizona, February 1992.

Davis G.A.: One Project, Two Discount Rates. Mining Engineering, April 1998.

Helfert E.A.: „Techniques of Financial Analysis – A Guide to Value Creation. 10th Edition, McGraw-Hill 2000.

Jankowska-Kłapkowska A.: Efektywność gospodarowania zasobami mineralnymi, Państwowe Wydawnictwo Ekonomiczne, Warszawa 1992.

Kasiewicz S., Rogowski W.: Analiza wrażliwości jako metoda analizy ryzyka przedsięwzięć inwestycyjnych.

Mat. Konf.: "Efektywność źródłem bogactwa narodów", 19–21 stycznia Wrocław–Karpacz, AE we Wrocławiu. Wrocław 2004, http://efektywnosc04.ae.wroc.pl/Referat/art24.pdf

Lawrence M.J.: DCF/NPV Modelling: Valuation Practice or Financial Engineering. SME Valuation Seminar, February 2000.

Mielcarz P.: Wykorzystanie narzędzi uzupełniających analizę zdyskontowanych przepływów pieniężnych netto w ocenie projektów badawczo-rozwojowych. Mat. Konf. "Efektywność źródłem bogactwa narodów", 19–21 stycznia Wrocław–Karpacz, AE im. Oskara Langego we Wrocławiu, Wrocław 2004, http://efektywnosc04.ae.wroc.pl/Referat/art52.pdf

Moyen N., Slade M., Uppal R.: Valuing Risk and Flexibility: A Comparision of Methods; Resources Policy, Vol. 22, Nos ½, 1996.

Pazio W.J.: Analiza finansowa i ocena efektywności projektów inwestycyjnych przedsiębiorstw, Oficyna Wydawnicza Politechniki Warszawskiej, wyd. II (popr. i rozszerzone), Warszawa 2002.

Runge I.C.: Mining Economics and Strategy, Society for Mining, Metallurgy, and Exploration, Inc., Littleton, CO 1998.

Simonsen H., Perry J.: Risk Identification, Assessment and Management in the Mining and Metallurgical Industries. The Journal of The South African Institute of Mining and Metallurgy, October-December 1999.

Smith L.D.: Discount Rates and Risk Assessment in Mineral Project Evaluations. Transactions Institution of Mining & Metallurgy (Sect. A: Mineral Industry) 1994.

Smith L.D.: Discounted Cash Flow Analysis and Discount Rates, Special Session on Valuation of Mineral Properties Mining Millennium 2000, March 8, 2000, Toronto, Canada, www.cim.org/mes/pdf/VALDAYLarrySmith.pdf, 2000.

Sobczyk E.J.: Praktyczne aspekty efektywnej eksploatacji złóż węgla kamiennego. Gospodarka Surowcami Mineralnymi, Tom 18, Zeszyt specjalny. Kraków 2002.

Szapiro T. (red.) + zespół: Decyzje menedzerskie z Excelem, PWE. Warszawa 2000.

Torries T.F.: Evaluating Mineral Projects: Applications and Misconceptions, Society for Mining, Metallurgy and Exploration, Inc. 1998.

Walls M.R., Eggert R.G.: Managerial Risk-Taking: A study of Mining CEOs. Mining Engineering, March 1996.

Wanielista K., Saługa P., Kicki J., Dzieża J., Jarosz J., Miłkowski R., Sobczyk E.J., Wirth H.: Wycena wartości zasobów złoża. Nowa strategia i metody wyceny. Wydawnictwo IGSMiE PAN, Biblioteka Szkoły Eksploatacji Podziemnej, Seria z Lampką Górniczą nr 12. Kraków 2002.

International Mining Forum 2005, Sobczyk & Kicki (eds) © 2005 Taylor & Francis Group, London, ISBN 0415 375525

Considerations, Construction and Influence of Tunneling with Small Overburden in Urban Areas

Motokazu Izawa
Taisei Corporation, Local Area Manager. Slovakia

Toshihiko Aoki
Taisei Corporation, Sitina Tunnel Project Manager. Slovakia

Daisuke Konno
Taisei Corporation, Sitina Tunnel Construction Manager. Slovakia

Pavel Zuzula
Taisei Corporation, Sitina Tunnel Engineer. Slovakia

Martin Sykora
Taisei Corporation, Sitina Tunnel Geologist. Slovakia

Marek Riesz
Taisei Corporation, Sitina Tunnel Geologist. Slovakia

ABSTRACT

When constructing tunnels in urban areas the influence of small overburden has to by often considered. Definition of small overburden is when its width is between 2 D–1,5 D or less. The influence can be divided into two parts. Above the tunnel and for the tunnel itself. Above the tunnels one has to pay attention to damage to buildings, roads or all kinds of utility lines, which can be damaged by surface subsidence or vibrations from blasting, and the influence of noise on inhabitants. For tunnel proper support patterns have to be used with an emphasis placed on an evaluation of the influence of small overburden and the support's efficiency has to be controlled by evaluation of geotechnical measurements. By considering these factors urban tunnels can be constructed properly and without negative influence on its neighborhood.

1. INTRODUCTION

The Sitina Tunnel project is located in Bratislava – the capital city of the Slovak Republic. This tunnel forms a part of the "Motorway D2 Lamacska Cesta – Stare Grunty" project. This is the last part of the motorway bypass around Bratislava and plays an important role for this region's traffic.

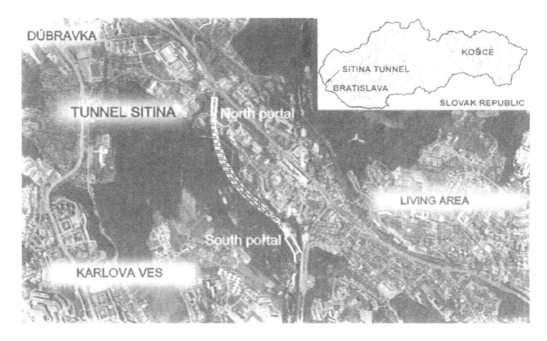

Picture 1. Location of Sitina Tunnel

The Sitina Tunnel is the first urban and two lane highway tunnel project in the Slovak Republic. We are also talking about first double-tube tunnel, which is being constructed by full profile, without a pilot tunnel. The tunnel consists of two tunnel tubes – East and West.

The tunnel is developed using the NATM method. Three most important aspect result from this fact. The tunnel is located in an urban area in a very close proximity to residential areas and the Slovak Academy of Science, the vibration and noise can influence the people and the structures close to the construction site. From the point of view of the construction small overburden (approx. 17 m) can be a problem.

Picture 2. View of South Portal

Picture 3. View of North Portal

2. GEOLOGICAL CONDITIONS

The Sitina Hill is situated in the western part of the Carpathian belt, in Male Karpaty Mountains. The massif is heavily broken by tectonic movements and fault zones (Vidrica fault and Lamačsky fault). The main directions of discontinuities in relation to the direction of the tunnel are perpendicularly 250/50 and parallel 130/70. The rock types are represented by Devon rocks (granodiorites, aplites and pegmatites), Neogen sediments and Quaternary sediments. No major underground water was expected from hydrogeological and geological study. All underground waters come from rain. Maximum expected inflow of underground water into the tunnel was 5 l/s. The average thickness of the overburden was around 18 m.

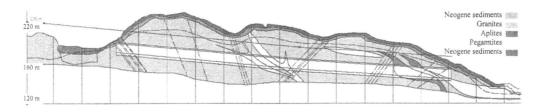

Picture 4. Geological cross-section of Sitina Tunnel

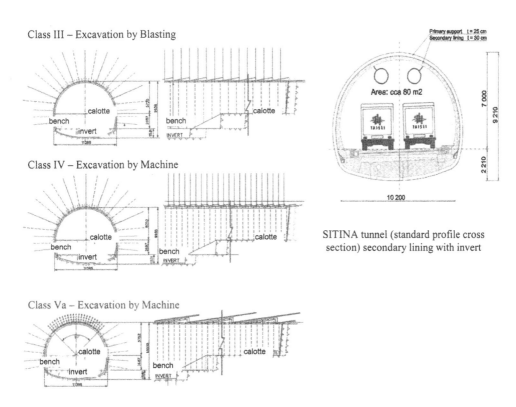

Picture 5. Typical primary lining cross-section for support classes III, IV, Va
and typical cross-section of secondary lining

155

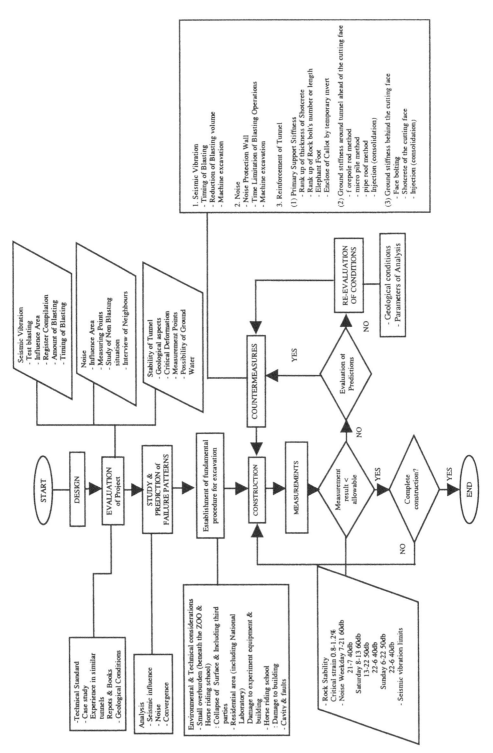

Chart 1. Project flowchart. Protocol of Urban Tunneling

3. SPECIFIC ASPECTS OF URBAN TUNNELS

Because of the location of the Sitina Tunnel some special aspects should be considered. Between the main considered aspects are Seismic monitoring, Pasportization, Noise measurements and Convergence measurements.

3.1. *Tunnel stability*

In urban districts where overburden width is particularly small, it is difficult to form ground arches and maintain the stability of the cutting face, so much support work has to be done. Numerous researchers have discussed the evaluation of the competence of surrounding ground as it concerns structural stability of tunnels. For example, (Tikaosa Tanimoto et al. 1984) formulated basic concepts for support design by investigating the spread of inelastic zones using ground competence factors, initial stress and support intensity that were derived from a strain-softening model. In addition, they cited the relation between convergence measurements and the incompetent region. (Tikaosa Tanimoto 1982; Tikaosa Tanimoto, Naoya Yoshioka 1991).

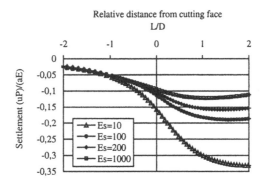

Figure 1. Effect of enhancing the support stiffness

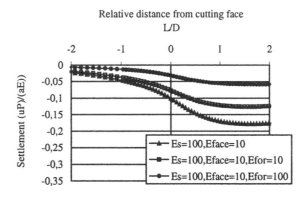

Figure 2. Effect of enhancing ground stiffness around tunnel periphery ahead of the cutting face

157

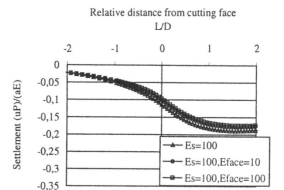

Figure 3. Effect of enhancing ground strength ahead of the cutting face

By considering the tunnel's stability, we can roughly divide the methods of enforcing it as follows:
- Figure 1 – Enhancement of support stiffness behind the cutting face.
- Figure 2 – Enhancement of ground strength around the tunnel periphery ahead of the cutting face.
- Figure 3 – Enhancement of ground strength ahead of the cutting face.

An analysis using finite element method was conducted with a three-dimensional axi-symmetric model. Based on these results, and examples of actual projects and measurements, we considered the effectiveness of each type of method.

According to the analysis, not only did the enhancement of support stiffness suppress convergence (the convergence behind the face stopped), it also played a major role in suppressing the preconsolidation convergence (the convergence ahead of the face stopped) (fig. 2, 3). However, when the rock at the cutting face was fragile, we noticed that even if support was installed near the perceived position of the face, the actual face (the solid ground ahead of the soft rock) acting as a part of the true support tended to be located in a more interior region than the perceived face. Thus, it appears to be extremely important to keep the cutting face function as a part of the support, which is known as the half dome effect of the face.

In this manner, in order to sustain the tunnel structure stability, we stressed the points to consider as below:
a) limitation of crown settlement (< critical stress 0.8–1.2%);
b) limitation of convergence (< 2D, D representative diameter of tunnel = 12 m);
c) preservation of cutting face function (technology such as shotocrete, face bolts, micropiles and forpole lods).

3.2. Seismic monitoring

Before starting the construction its influence area has to be defined. All objects situated in this area have to be characterized with regard to stability, resistance to vibrations and influence of vibrations on people. After evaluation of all these objects, initial register of endangered object has to be done.

If there is a possibility of damage to surrounding objects from vibrations, seismic monitoring has to be executed. Seismic monitoring contains of the following steps.
1. Classification of endangered buildings in the influence area.
2. Executing trial blasting to define rock properties.
3. Installing seismographs to monitor elastic velocity of vibrations in specified places around the tunnel. Replacing the seismographs accordingly to the tunnel's advance.

4. Continuous evaluation of measurements (data receiving daily by e-mail) and correction of blasting works in the tunnel to minimize the influence and to avoid damages to nearby objects.

The Sitina Tunnel is situated near the Slovak Academy where many sensitive measuring instruments are installed. For this reason the influence to this equipment was evaluated in more detail. Also 30 m above the tunnel a horse riding school is situated. To minimize the influence of vibration on people and animals changes to the design of the blasting works had to be done in this place.

3.3. *Noise monitoring*

Because of the noise from the construction site caused by blasting works, ventilation fans and machinery noise measurements had to be done. Noise monitoring contains of the following steps:
1. Classification of the most sensitive parts of the neighborhood.
2. The values of the highest admissible noise levels are defined for day-time (6:00 to 22:00) and night-time (22:00 to 6:00). A reference quantity for noise measuring is the equivalent noise level. Government Order of the Slovak Republic specifies the highest admissible values of the equivalent noise levels for noise outdoors (applicable to Sitina Tunnel).
3. Measuring the noise level in the sensitive places of the neighborhood.
4. Evaluation of noise levels.
5. Design of noise reduction measures.
6. Measuring the noise level in the sensitive places of the neighborhood.
7. Evaluation of the efficiency of the noise reduction measures.

Protection is provided by anti-noise doors in booth tunnels and anti-noise barriers of fans. Noise monitoring checked efficiency of these protection measures. Measuring points were situated in front of the South portal in a block of flats (500 m), a gas station (300 m), two Family houses (200 m), School Rosa (500 m) and Pneuservis (120 m). From the result changes to the blasting works were designed so that the levels did not exceed the Government regulations and blasting could be executed in a 24 hrs cycle.

3.4. *Convergence measurements*

In urban tunnels with small overburden support patterns can be evaluated according convergence measurement.

4. RESULTS

4.1. *Seismic monitoring*

From the results of seismic monitoring the influence to the surroundings can be defined (fig. 4).

4.2. *Noise measurements*

Purpose

With respect to the works on Sitina Tunnel, there is high possibility for an unwanted interference of interest to be involved, since the building site (including the tunnel part) is situated in densely populated area. Therefore, to know the effect of building works on the environment is very important.

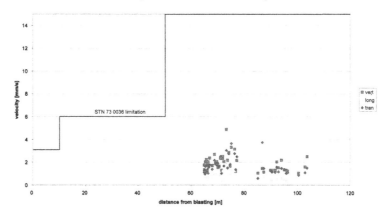

Figure 4. The influence to the surroundings from the results of seismic monitoring

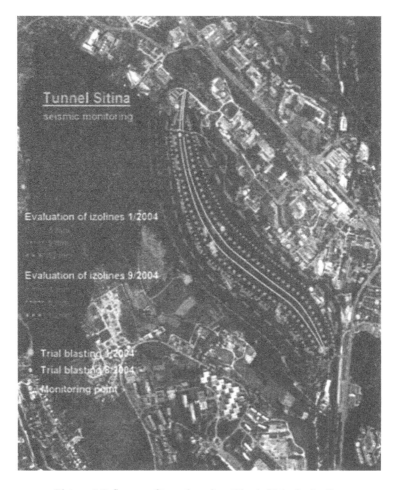

Picture 4. Influence of tunnel works – Elastic Velocity Isolines

160

Protocol

Values of highest admissible value are defined for daytime (6:00 to 22:00) and night-time (22:00 to 6:00). A reference quantity for measuring noise is the equivalent noise level[1].

Government Order of the Slovak Republic[2] specifies highest admissible values of equivalent noise level for noise outdoors (applicable to Sitina Tunnel):
- In day-time the limit value is 50 dB.
- In night-time the limit value is 40 dB.

During construction activities on weekdays from 7:00 am to 9:00 pm and Saturdays from 8:00 to 1:00 pm correction – 10 dB from measured value is used. Also, depending on the character of sound different corrections[1] has to be used.

Results

First results from measuring the noise from blasting and ventilation fans were not positive. The values exceeded the limits specified in the Government order.

Values obtained from the next measurement for day- and nighttime for times with and without blasting are very similar, as seen in figure nr.1. After installation of noise doors the equivalent noise level was decreased by 6 dB. As the distance from the portal increases, the noise level is lower (for distance 185 m the noise decreased by approximately 10 dB, for 15–20 kg of explosives, see figure 5). Using new ventilation fans decreased the noise level by 25 dB.

It was due to the combination of these countermeasures that the level of noise was decreased. One of the important factors is also timing of blasts in partial charges, which changes the character of the sound. Finally, the contribution of blasting to the already existing noise from traffic is minimal and hardly interpretable.

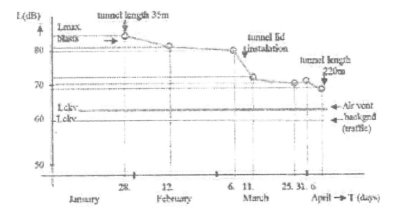

Figure 5

EQUIVALENT NOISE LEVEL (dB)	DAYTIME	NIGHTTIME
Traffic noise	60.7	54.4
Noise from blasting	61.3	54.2

---

[1] STN ISO 1996-2, STN ISO 1996-2/Amd.1.
[2] Government Order of Slovak Republic Nr.40 from 16/1/2002: "About Protection of Health against the Noise and Vibrations".

4.3. *Convergence measurements*

From the results of convergence measurements the time when the stabilization of deformation occurred can be evaluated. If until 2D, then the support class is efficient. If more than the 2D the support class is not efficient. In graphs (fig. 6) the evaluation of measurements for both tunnel tubes is shown.

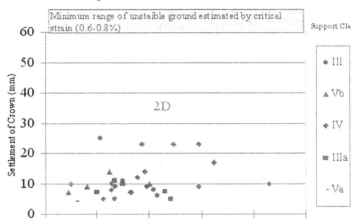

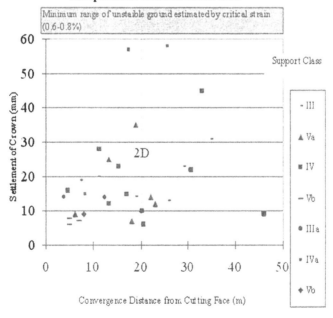

Figure 6. Evaluation of convergence measurements in West and East Tunnel Tube

CONCLUSION

By considering all interpretations of seismic monitoring, noise measurements and convergence measurements, construction of urban tunnels can be properly evaluated. Based on these evaluations we can solve the problems associated with the construction.

Picture 5. Inside and Outside of the tunnel

REFERENCES

Tikaosa Tanimoto: NATM-1. Morikita Shuppan Co. Ltd. 1984.
Tikaosa Tanimoto: NATM as an Observational Construction. Soil and Foundations, Vol. 30, No 7, pp. 63–70, 1982.
Tikaosa Tanimoto & Naoya Yoshioka: Measuring of Convergence Measurement in Tunneling. Extra Issue of Materials Science, Vol. 40, pp. 630–636, 1991.

International Mining Forum 2005, Sobczyk & Kicki (eds) © 2005 Taylor & Francis Group, London, ISBN 0415 375525

Use of Mineral Coal for Sorption Sewage Treatment

A.V. Mozolkova
Russian University of People's Friendship. Moscow, Russia

E.V.Chekushina
Russian University of People's Friendship. Moscow, Russia

A.A. Kaminskaya
Russian University of People's Friendship. Moscow, Russia

Treatment of mining, industrial, household and other sewage is an actual problem for many mining and processing enterprises. Coal-mining industry is not an exception. Usually, at coal enterprises, treatment of mine sewage before it is dumped consists in settling and subsequent filtering. Many pollutants are not removed from the sewage by this method. Hence, dumped water frequently does not satisfy sanitary requirements regarding the permissible content of oil products, dissolved substances and other parameters.

For additional cleaning of sewage it is possible to use sorption methods. By these methods water is cleaned of oil products, heavy metals, a number of organic substances and other polluting substances, depending on the used sorbent properties. Both natural and artificial materials can be used as sorbents. Constraint for wide use of sorption methods of sewage treatment in the coal industry is high cost of the majority of sorbents.

A number of technologies for obtaining inexpensive and good quality sorbents from coal minerals have been developed. These sorbents can be manufactured directly in coal mines which has additional advantage of reducing transport costs. The processed sorbents may be recycled or burnt. Apart from that production and sale of sorbents can serve the coal-mining enterprises as an additional source of income.

One of the most widespread sorbents is activated coal. Quality activated coals are carbon sorbents, having an internal specific surface of more than 500 m^2/g, and characterised by iodine adsorption (iodine value) of more than 500 mg/g. Mineral coal, peat and wood can serve as raw materials for activated coal production. Traditional production techniques of activated coal include two basic stages of thermal processing of the initial carbon-containing raw material – carbonization and activation, done in different devices. Both stages are energy-consuming and ecologically dangerous, which explains the high cost of activated coal, received through this technology (1200–4000 dollars/MT).

Carbonization is the elimination of volatile substances by heating raw material up to the temperature of 600–900°C, because with volatile components there are basically formed the oxygen and hydrogen, and increase carbon content in initial raw material. Carbonization is done in mining or rotating furnaces with utilization of external form-holder, as a rule, waste gases with temperatures of 600°C and higher.

Activation means increasing the volume and pore surfaces of carbonized material at heterogeneous reaction. The most used reagent is water vapour with the temperature of 900°C and higher, and the process takes 15–20 hours.

Both stages are energy consuming and pose threat to environment. For one tonne of activated coal from 2 up to 4 tonnes of specific fuel like crude oil and natural gas are consumed. From 1000 up to

1500 m³ of processed gasses with high content of SO_x (1–2 g/m³), H_2S (200–250 mg/m³), resinous substances (10–40 mg/m³), phenols (50–70 mg/m³), carbon oxides (up to 5%) and also other substances which are carcinogenic and mutagenic are formed and released into the atmosphere during both stages. High-energy consumption and environmental danger, which requires large investments in nature protection activities, result in the high cost of quality-activated coal.

Another group of carbon sorbents, which was widespread in the 80's, consists of inexpensive carbon sorbents used in nature protection technologies and industry. Such sorbents are produced by a one-phase technology, without additional activation. Their adsorption activity is not high (iodine value less than 300 mg/g) but the cost is low (250–700 USD/MT). Because the price of these sorbents is comparable to the cost of their regeneration, they are used only once and are burnt after saturation. The leaders in the production of such sorbents are Rheinbraun AG (Germany, 200 thousand MT per year) and Australian Char Pty Ltd (Australia, 150 thousand MT per year), which produce brown coal semi-coke used for treatment of wastewater and smoke. In Russia research work in this direction is conducted, but only test works have been done so far, although the quality of carbon sorbents obtained from Kansk-Achinsk coals did not concede to production from Rheinbraun AG and Australian Char. One of the directions of utilization of semi-coke from Kansk-Achinsk coals, production of which was planed at Krasnoyarsc thermal power station 2 (device ETX-175), was its utilization as carbon sorbent.

The reason for production of inexpensive carbon sorbents by one-phase technologies being not developed in Russia is the absence of demand for this product. This production is basically used for cleaning of sewage, however there is no effective ecological service in Russia, and the penal sanctions of the environmental protection legislation are so insignificant that industrial enterprises do not have any motivation to invest in nature protection.

In 1992–1994 the employees from Joint-Stock Company "Carbonica-F" (at that time Open Company "Sibtermo") have developed and have patented a new method of production of carbon sorbents, which considerably differed from all known technologies.

During the research of dynamic effects in a layer evaporator the regime conditions were defined under which the effect of "thermal wave" could be observed in the device. Using this effect, the authors created a layer evaporator in which volatile components of coal were exposed to gasification (incomplete oxidation), and the degree of carbon conversion was adjusted by the mode of injection feed. By changing the regime parameters it was possible to conduct the process as fuel gasification (with only ashes remaining in the end) without any residue, and also as gasification of volatile components of coal, thus receiving so-called semi-coke containing solid coal. From one tonne of Kansk-Achinsk coal with calorific content of 3600–3800 kcal/kg can be produced about 0,33 tonnes of semi-coke with calorific content up to 7000 kcal/kg (as anthracite) and up to 1700 m³ of combustible gas with calorific content of 800–900 kcal/m3, suitable for use as an energy source.

Technological process of Joint-Stock Company "Carbonika-F" has a large number of advantages in comparison to the already known methods of obtaining activated coal and semi-coke:

1. Simplicity of hardware. One-phase process. The stages of drying, pyrolysis, thermal decomposition of volatile substances and semi-coke cooling are incorporated in one device. The device is auto- metric; it means that external heat-carrier for coal heating is not used.

2. Ecological safety. In the technology of Joint-Stock Company "Carbonica-F" all hydrocarbons, including resinous substances, are broken down and gasified inside the device during the formation of combustible gas containing only CO, H_2, CO_2, N_2, H_2O, H_2S and insignificant quantity CH_4. Sludge, pyroligneous waters, phenols and other harmful impurities are not formed in this process.

3. Because the speed of gas filtration from a layer reactor is low (0,02–0,03 m/s in comparison to 0,5–2,5 m/s for mine furnaces), the process is less dependent on fractional composition of coal, hydraulic resistance of the layer and allows to process fine-grained coals.

4. As a result of low speed of filtration the phenomenon of carrying out of fly ashes from the layer does not occur, because the device works as a granular filter. Combustible gas is moved in user-boiler or can move to the gas turbine without preliminary cleaning. The volume of SOx, NOx, CO contained in waste gases is lower than that produced when obtaining equivalent quantity of heat by burning coal. Combustible gas without prior cleaning can be used to produce electric and/or thermal power or as an energy carrier for thermal processes.

5. Unlike the already existing technologies, in the given process there is no dump of gaseous heat-carrier into the atmosphere and consequently the construction of other additional gas purification systems and catalytic burning of carbon oxide (CO) is not required.

Test of the solid residue (semi-coke) have revealed, that this material is characterized by large specific surface (more than 500 m^2/g) and high adsorption activity (iodine value 500 mg/g and higher), and because of these parameters does not concede to quality-activated coal.

The product received with the technology of Joint-Stock Company "Carbonica-F" is certificated as activated coal ABG (active, brown coal of gasification), for it there were developed technical conditions TU 6-00209591-443-95. The characteristics of ABG activated coal produced from the coal of b2 mark from "Berezovsky-1" opencast colliery, are shown in table 1.

High specific surface and adsorption activity of ABG coal is explained by the fact that both gasification of coal volatile components, and activation of carbon-containing solid residue of gaseous products occur in the device simultaneously. Because gasification products contain up to 20% of hydrogen whose molecules are smaller than the ones of water vapour, and hence their permeability in pores of semi-coke is higher, activation (heterogeneous reaction) is done not only with vapour, but also with hydrogen, which practically is not present in the traditional technologies. Thus, carbonisation stages and activation are combined in one device.

Other positive effect of application of this method of coal processing is that in "thermal wave" mode the products of thermal decomposition which contain very toxic resinous substances (coal tar pitch used in experimental medicine for the inoculation of cancer in experiments on mice, brown coal is more toxic), passing through a hot layer of semi-coke (500–700°C) are completely broken down into two and three-nuclei gases H_2O, CO_2, CO, H_2. Measurements done at the working production plant of Joint-Stock Company "Carbonica-F" have shown that the gas does not contain hydrocarbons of lines above methane, and also carcinogens, including benzo(a)pirene.

Table 1. Characteristics of ABG activated coal produced from the coal of B2 mark from "Berezovsky-1" open cast colliery

PARAMETER	NOTATION	MEASUREMENT UNIT	VALUE
Data of technical analysis			
General moisture	W^r_t	%	2
Ash content	A^r	%	up to 10 (12)
Volatiles content	V^r	%	7–9
Strength at friction	P	%	70
Bulk density	ρ_{NAS}	g/cm^3	0,45–0,5
Adsorption properties			
Iodine adsorption activity		%	60
Adsorption activity of methyl blue		%	95
Parameters of porous structure			
Specific surface	S_Σ	m^2/g	500
Total pores volume	V_Σ	cm^3/g	0,57
Volume of micro pores	V_{mi}	cm^3/g	0,2
Volume of mesopores	V_{me}	cm^3/g	0,25
Volume of macro pores	V_{ma}	cm^3/g	0,12

Cooling of the activated coal from 550 up to 70°C before discharging is carried out by compulsory circulation of gaseous heat-carrier (waste gases) through a layer of the product and further through shell-and-tube heat exchanger in which water used in closed circuit is also provided. Total process efficiency reaches 95% due to the high degree of utilization, which is associated with utilizing the thermal energy.

Departing waste gases do not undergo any cleaning; there are even no cyclones. Nevertheless, the content of harmful mixtures (NO_x 150 mg/m^3, SO_x 50 mg/m^3, ash less than 10 mg/m^3) is essentially lower than the established norms and parameters of working boiler and thermal power stations, even those equipped with modern multistage systems of gas purification including electro filters. This is explained by a insignificant ablation of ash from devices, sorption of sulfur compounds in activated corner, and also focus temperature from the user-boiler is lower than 1600°C – "threshold" temperature at which begins the formation of nitrogen oxides due to the oxidation of nitrogen from the air.

The technology of Joint-Stock Company "Carbonica-F" can be used for any not conglomerating coals.

Similar sorbents or slightly conceding in quality to activated coal are formed by semi-coking of unconglomerated coal. Semi-coke received by using the technology developed and patented at Joint-Stock Company "Carbonika-F" is characterised by large specific surface (above 500 m^2/g) and high adsorption activity (iodine value 500 mg/g and more), and with these parameters does not concede in quality to activated coal. The production of this sorbent is ecologically safe. The production by-product – combustible gas can be burnt in boilers of thermal power station.

Some mineral coals (called mesoporous) have internal pores accessible to water, having the size 3,5−4 nanometers (mesopores), forming active surface, sized 50−120 m^2/g (unlike all other natural coals with surface of 0,5−1 m^2/g). These coals can be used as sorbents without additional activation. They clear water of undissolved and dissolved mineral oil, deep dispersing mixtures, iron, phenol, ions of heavy metals, ammonia, nitrates, benzo(a)pirene and so forth. Sorbent MIU-S received from poorly metamorphosed mesopore coal can be used for 3−7 years with periodic regeneration. Alkali regeneration solution is removed from the filter without other additional neutralization, because in alkali and acid medium MIU-S presents buffer properties, neutralizing these media.

Specific porous structure of mesopore coals assures sorption extraction of dissolved mineral oil products with concentration lower than 1 mg/l, and thus is not always reachable even with activated coals.

Using MIU-S filters in drinking water supply systems made the stability of their work in conditions of continuous exploitation evident, maintaining the properties of sorbents at null and sub-zero temperatures and absence of biomass formation.

Besides the abovementioned technologies, sorbents can be obtained from mineral coal by its briquetting and activation. Raw material for briquettes can be coals of any rank.

Thus, sorbents suitable for additional cleaning of sewage are possible to be produced from mineral coals by special processing, and sometimes directly. Production of own sorbents may solve the problem of additional cleaning of sewage in coal enterprises. Mesopore coals can be used as sorbents without additional processing; the other coals need additional activation. The studied sorbents can be used for cleaning sewage water from mineral oil products, organic substances, and metal ions.

REFERENCES

Kovaleva I.B., Matvienko N.G., Solovyeva E.A., Tarnopolskaya M.G.: The Application of Natural Mineral Coal in the Technology of Sewage Treatment from Mineral Oil. World Congress on Mining Ecology. Works of the Congress 1999, pg. 310–315.2.
For the preparation of the article have been used materials from the site www.carbonica.scn.ru, www.miu-sorb.ru

International Mining Forum 2005, Sobczyk & Kicki (eds) © 2005 Taylor & Francis Group, London, ISBN 0415 375525

The Influence of Microstructure and Mineral Ingredients in Coal on Its Spontaneous Combustion

He Qilin
Anhui University of Science & Technology. Huainan , Anhui, China

Yuan Shujie
Anhui University of Science & Technology. Huainan, Anhui, China

Wang Deming
China University of Mining & Technology. Xuzhou, Jiangsu, China

ABSTRACT: For different ranks of coal or coal with the same coalification degree but different susceptibility to spontaneous combustion, the S250MK3 scanning electron microscope was used at different amplifications to observe their overall appearance, pores and fracture distribution. The observation by the S250MK3 scanning electron microscope combined with the mineral composition ascertained by the mineral substances energy spectrum analysis elicited the conclusions: coal with different coalification degree has significantly different porous structure; the porous structure of coal is the main factor determining the quantity of oxygen absorption by coal at low temperatures, which has certain effect on susceptibility of coal to spontaneous combustion. The research also found, that sulfur content and its different occurrence in coal is the main cause resulting in different susceptibility to spontaneous combustion of coal with the same coalification degree.

KEYWORDS: Scanning electron microscope, microstructure of coal, oxygen absorption, spontaneous combustion susceptibility

1. INTRODUCTION

Coal was formed from plant matter and dead animals by physical, chemical and biological processes. Its formation went through quite long geological periods. The remains of paleo-plants and animals gradually formed this organic rock fuel under the effect of high temperature and pressure (Xie 2002). Because of many factors, such as complexity and diversity of paleo-organism, different natural conditions during coal formation and different coal formation periods, there are differences in coal structure, the status of pores developed in coal, mineral ingredients filled in fractures, etc. Therefore, different ranks of coal have different oxygen absorption capacity, oxygen absorption rate and heat release during coal oxidation. And further more, different ranks of coal or the same rank of coal with different microstructure or mineral ingredients may possibly have different susceptibility to spontaneous combustion.

In order to study the influences of different porous structures in different ranks of coal on its spontaneous combustion, the S250MK3 scanning electron microscope made in Britain was used at different amplifications to observe the overall appearance, pores and fractures in coal. Different ranks of coal (lignite, bituminous coal and anthracite) and coal with the same coalification degree but with different susceptibility to spontaneous combustion (samples from Datun Kongzhuang Colliery of Datun Coal & Energy Company and Chaili Colliery of Zaozhuang Coal Company) were examined in the Analysis and Measurement Center at the Chinese University of Mining and Technology in Xuzhou. The mineral substances energy spectrum was measured to analyze the composition of mineral substances filling the pores (Tooke 1983). By the method, we expected to find out what influence microstructure of coal affects on its susceptibility to spontaneous combustion.

2. METHOD OF EXPERIMENT

At first, coal samples were worked manually into standard size and shape. The selected coal samples to be processed should have moderate thickness and fresh fracture section along the layer side. According to the demands of the experiment, coal samples should be processed into cylinder cores with a diameter of 1 cm and thickness of 5 mm. Then latex (polyvinyl acetate) was glued onto the coal samples. At last, a coating of gold 0.01~0.1 μm thick was plated on the coal samples with the ion sputtering technology. The gold coating adhered to the latex to ensure good contact with the coal samples.

When the coal samples had been processed, they were placed into the sample chamber and kept in a vacuum for half an hour. Then the overall appearance of coal samples was observed at low amplification and their microstructure observed at gradually increased amplifications.

3. RESULTS AND DISCUSSION

3.1. *The influence of microstructure of coal on its oxygen absorption at low temperatures*

Figure 1 shows the comparison of overall appearance of lignite and anthracite and figure 2 the comparison of the state of pores and fractures developed in coal. From figure 1 and figure 2, some features of lignite, bituminous coal and anthracite can be seen. Lignite has a very loose structure and a very high degree of pulverization. A lot of loosely bound conglomerates and granules having irregular shapes and different sizes are clearly seen. The overall structure of the coal samples shows powder and floc appearance. Pores and fractures developed very well. Mineral substances occur in the way of heaping each other or infiltrating into and integrating with coal. Bituminous coal has a more compact structure, better integrity and more compact microstructure than lignite. In the samples tiny pores are intensely distributed and in some parts of the samples porous structure is seen. Pores with diameter smaller than 1 μm dominate. The pores split and formed fractures, generally without any matter filling them. In some parts of the samples there are pores with diameters smaller than 5 μm. Tiny pores form a porous network, having good interconnections. Fractures with diameters smaller than 1 μm are fully developed. Fractures interconnect with each other and are apt to be filled with mineral matter. Anthracite has a compact, solid structure with good integrity and its overall fracture section is very flat and smooth. Pores with diameters bigger than 5 μm predominate and tiny pores are not well developed. Pores have bad interconnections and are filled with mineral matter in the way of ridgelike, inlaid structure. Inlaying bodies closely combine themselves with each other. Fractures occur because of coal seam cracking and are generally filled with mineral substances.

From the results obtained by scanning electron microscope measurements it we can be learnt that different ranks of coal are characterised by great differences of microstructure. Therefore, different structural surface areas in different ranks of coal result in differences in oxygen absorption capacity and heat release. Microstructure in different ranks of coal is the main factor affecting the capacity of oxygen absorption at low temperatures.

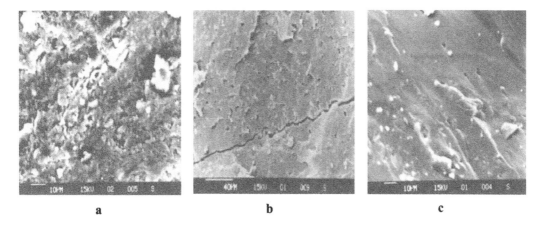

a   b   c

Figure 1. Comparison of overall appearances of different coals:
a) Lignite from Baizao Colliery, Longkou Coal Company;
b) Bituminous coal from Kongzhuang Colliery (seam 8), Datun Coal & Energy Company;
c) Anthracite from Bbaishan Colliery, Huaibei Coal Company

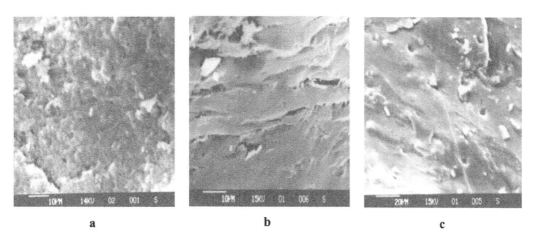

a   b   c

Figure 2. Comparison of the state of pores and fractures developed in different coals:
a) Lignite from Baizao Colliery, Longkou Coal Company;
b) Bituminous coal from Kongzhuang Colliery (seam 8), Datun Coal & Energy Company;
c) Anthracite from Bbaishan Colliery, Huaibei Coal Company

3.2. *The influence of sulfur in coal on its susceptibility to spontaneous combustion*

Mineral substances contained in coal are called inorganic micro-ingredients. Their sources include inorganic ingredients contained in plants forming coal and some minerals interlarded during coal formation. The latter is the main source of mineral substances in coal. Some most common minerals are mainly clay minerals, sulfide minerals, oxides and carbonates etc. The main component substances of sulfur in coal are pyrite and its oxidation product copperas ($FeSO_4 \cdot 7H_2O$), which release heat, when oxidizing in humid air. The formulas of the chemical reactions are as follows:

$$2FeS_2 + 2H_2O + 7O_2 \rightarrow 2FeSO_4 + 2H_2SO_4 + Q_1 \tag{1}$$

171

$$12FeSO_4 + 6H_2O + 3O_2 \rightarrow 4Fe_2(SO_4)_3 + 4Fe(OH)_3 + Q_2 \qquad (2)$$

$$FeS_2 + Fe_2(SO_4)_3 + 2H_2O + 3O_2 \rightarrow 3FeSO_4 + 2H_2SO_4 + Q_3 \qquad (3)$$

$$FeS_2 + O_2 \rightarrow FeSO_4 + SO_2 + Q_4 \qquad (4)$$

The reactions mentioned above are exothermic. Where Q_1, Q_2, Q_3 and Q_4 in the formulas represent heat releases.

Some basic parameters of the studied coal samples are given in table 1. From the table it can be seen that even though the coal samples from Baishan Colliery of Huaibei Coal Company and No. 5 Colliery of Yangquan Coal Company are both anthracite, the actual susceptibility to spontaneous combustion of the two coal beds is vastly different because of their different sulfur content. In Baishan Colliery of Huaibei Coal Company spontaneous combustion never occurred. However, in No. 5 Colliery of Yangquan Coal Company the shortest period of incubation is only thirteen days.

Table 1. Basic parameters of coal samples

Samples	Actual incubation period (month)	Industrial analysis (%)				Calorific value (J/g)	Element analysis (%)				
		M_{ad}	A_{ad}	V_{ad}	FC_{ad}	$Q_{net, ad}$	C_{ad}	H_{ad}	N_{ad}	$S_{t, ad}$	O_{ad}
Bituminous coal from Kongzhuang Colliery of Datun	4	1.52	5.75	35.81	56.83	28674	76.27	4.58	1.33	0.52	10.04
Bituminous coal from Chaili Colliery of Zhaozhuang	1.5	1.51	6.15	37.36	54.98	27656	75.55	4.73	1.29	0.63	10.15
Anthracite from Baishan Colliery of Huaibei	Never firing	1.63	6.20	8.59	83.58	33500	86.12	3.34	1.37	0.27	1.07
Lignite from Baizao Colliery of Longkou	0.6	25.05	2..91	30.41	41.63	22236	54.92	2.96	1.63	0.37	12.16
Anthracite from No. 5 Colliery of Yangquan	0.43	1.73	6..31	8.26	83.70	32678	85.13	3.28	1.26	2.50	1.13

3.3. *The influence of differences in mineral content of coal on its susceptibility to spontaneous combustion*

Coal samples from Kongzhuang Colliery of Datun Coal Company and Chaili Colliery of Zaozhuang Coal Company have the same coalification degree. The content of the organic ingredients C_{ad} H_{ad} O_{ad} and inorganic elements, $S_{t, ad}$ in the coal samples are almost the same. From table 1 can be seen that during mining, the actual incubation periods of the coal samples are respectively 4 and 1.5 months. Because the coal beds actually have the shortest periods of incubation among all the

172

studied coal beds, namely, they have the best natural conditions for their oxidation. It can be supposed that differences in the mining technology utilized in different collieries make almost no difference as a technologic condition affecting spontaneous combustion. Why do coals with the same degree of coalification have different actual shortest periods of incubation? By observing carefully the two samples and measuring the energy spectrum of the mineral substances contained in them it is found that in the coal samples from Kongzhuang Colliery of Datun Coal & Energy Company, inorganic mineral substances are of very small size and are spread evenly in the coal. Mineral substances are closely integrated into coal, and the substances contained in the fractures and pores are mainly clay minerals and carbonates. In the coal samples from Chaili Colliery of Zaozhuang Coal Company, the substances present in fractures and pores are not only clay minerals and carbonates, but also pyrite in large quantities. The mineral substances of bigger size filling the ractures and pores show a tendency of heaping together. The mineral substances made the surrounding coal body crack and produce more interconnected fracturing, therefore becoming incompact (figure 3 and figure 4). During mining of such kind of coal, because of well developed fractures and good conditions for oxygen supply and accumulation of heat released during oxidation of carbonates, some localized high temperature points easily occur and result in spontaneous combustion of coal.

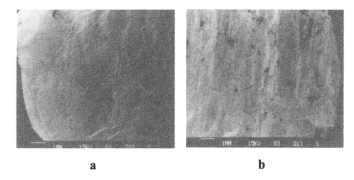

a   b

Figure 3. The comparison of overall appearances between coal from Kongzhuang and Chaili Collieries:
a) The overall appearance of coal from Kongzhuang Colliery;
b) The overall appearance of coal from Chaili Colliery

a   b

Figure 4. Comparison of the mineral substances accumulated in pores and fractures in coal
from Kongzhuang and Chaili Collieries:
a) The mineral substances accumulated in fractures in coal from Kongzhuang Colliery;
b) The mineral substances accumulated in fractures in coal from Chaili Colliery

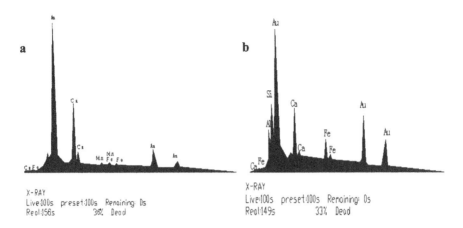

Figure 5. Comparison of the minerals accumulated in fractures and pores in coal samples
from Kongzhuang and Chaili Collieries:
a) The energy spectrum of the mineral substances accumulated in fractures in coal
from Kongzhuang Colliery;
b) The energy spectrum of the mineral substances accumulated in fractures in bituminous coal
from Chaili Colliery

4. CONCLUSIONS

1. The degree of coalification determines organic components and their spatial distribution in coal. It's an important factor affecting susceptibility of coal to spontaneous combustion. In the processes of coal formation from peat, lignite, bituminous coal to anthracite, because of the effect of temperature and pressure coal underwent successive processes of decarboxylation, dehydroxylation, dealkylation and dehydrogenation. The processes made the carbon content of coal and condensed-nuclei ratio of volatiles increase successively. Therefore, the structure of coal became more compact and more stable. The possibility of heat release due to coal oxidation decreases, and as a result the susceptibility of coal to spontaneous combustion decreases.
2. Sulfur content of microscopic inorganic components and its different occurrence in coal are the main factors resulting in different susceptibility of coal of the same rank to spontaneous combustion. The sulfur content in coal determines the capacity of heat release due to coal oxidation to take place and the manner of sulfur occurrence in coal determines the way of heat release: spread uniformly or localized in points. Coal with incompact heap structure of sulfur is apt to spontaneous combustion.
3. Observing the microstructure by scanning with an electron microscope and determining the mineral components in coal by an energy spectrum analysis helped to study the differences of microstructure of coal and the cause for different susceptibility of coal to spontaneous combustion. The study presented above is only qualitative. In order to determine the extent of the susceptibility of coal to spontaneous combustion, some parameters, such as quantity of oxygen absorption, oxygen absorption rate, quantity of heat released during coal oxidation must be measured at different temperatures.

REFERENCES

Xie K. 2002: Coal Structure and Its Chemical Reactivity [M]. Beijing Science & Technology Press.
Tooke P.B. and Grint A. 1983: Fourier Transform Infra-red Studies of Coal [J]. Fuel, Vol. 62, 1983 (9),pp. 1003–1008.

International Mining Forum 2005, Sobczyk & Kicki (eds) © 2005 Taylor & Francis Group, London, ISBN 0415 375525

Protection of the Atmosphere During Mining of Coal Deposits by Open Cast Methods

A.E. Vorobiev
Russian University of People's Friendship. Moscow, Russia

A.V. Mozolkova
Russian University of People's Friendship. Moscow, Russia

E.V. Chekushina
Russian University of People's Friendship. Moscow, Russia

A.A. Kaminskaya
Russian University of People's Friendship. Moscow, Russia

The structure of the Earth's atmosphere, established in the course of millions of years, is changing now under the influence of natural and anthropogenic factors. A significant contribution to atmosphere pollution is brought by the mining industry and especially by the extraction of minerals by open cast methods. In the mid-80's of the last century "Nerungri" open cast colliery released into the atmosphere about 10 000 tons of dust that comprised about 0,1% of its production.

During coal mining dust and toxic gases are released.

The sources of dust are the processes of coal preparation and stripping of overburden rock, excavation, transport, warehousing and settling, i.e. all technological processes. Due to its content of free silicon dust of many coal-containing rocks are regarded as silicosis risky, and coal dust, unlike rocks' dust, as a rule is not silicosis risky.

Usually dust elimination during coal exploitation is several times mofe efficient than during stripping of the overburden. Much larger quantities of dust are eliminated during coal mining at subzero temperatures. The reason is that the moisture which is contained in coal freezes and cannot combine with dust, besides it occurs its frost and that stronger diminishes the forces, which hold separate dust particles from passing into free state.

Inhalation of dust can cause specific diseases – pulmonary mycosis, contribution to contamination with these diseases, as laryngitis, tracheitis, bronchitis, lung tuberculosis and skin disease.

In coal industry there are also widespread such diseases of dust etiology as silicosis (the most serious illness) which appears as the result of action of coal dust with high content of silicon dioxide, antrocosis at influence of coal dust and antrocosilicosis at influence of coal-and-rock dust.

Dust can be the reason of accidents, it worsens working conditions of employees, reduces productiveness, contributes to fast wearing out of production installations.

In Russia gravimetric concentration of all dust inhaled in working zone is measured and normalized, in other countries, (except for CIA countries) – the gravimetric concentration of inhalable dust fraction (less than 5 microns).

The harmfulness of coal and rock dust depends mainly on the content of free silicon dioxide (SiO_2). The more silicon dioxide there is, the bigger is its toxicity. Toxic gases are released during blasting works and by diesel engines with which open cast collieries are equipped, except for that rocks, coal and underground waters can also be regarded as sources of gas release.

Dust concentration in the atmosphere of open cast colliery can exceed hundreds of mg/m^3, although the maximum permissible values for a working zone in Russia are from 1 up to 10 mg/m^3.

The intensity of dust formation and dust separation depends on wind speed, air temperature, humidity and temperature of processed coal and rock, dust prevention measures taken in the open cast colliery, physico-mechanical properties of the processed material.

Limits for maximum concentration of dust in the air has been established by sanitary norms. Table 1 presents maximum allowed concentrations of some dusts. In total more than 30 different types of non-toxic dust are normalized.

Table 1. Maximum allowed concentrations of dust of some substances in the atmosphere

SUBSTANCE	MAC, mg/m^3
ATMOSPHERE OF WORKING ZONE	
DUST CONTAINING SIO$_2$	
For the content of more than 70% of SiO$_2$ (quartz, cristoballite, tridimite) in dust	1
For the content from 70 up to 10% of SiO$_2$ (granite, shamotte, crude mica, carbonic dust etc.)	2
For the content from 2 up to 10% of SiO$_2$ (oil shale, coppersulphyde ores etc.)	4
CARBONIC DUSTS	
Anthracite containing free silicon dioxide up to 5%	6
Other mineral coals and carbonic dusts with the content of silicone dioxide up to 5%	10
ATMOSPHERE OF INHABITED PLACES (FOR NEUTRAL ATMOSPHERIC DUST)	
Daily average	0,15
Maximum-single	0,5

It can be said from experience that concentration of dust in the atmosphere of open cast collieries exceeds hundreds of mg/m^3. Usually the biggest dust particles settle down without leaving the limits of the works, but some part of the dust is thrown out beyond its limits and pollutes the atmosphere of adjacent territories, sets on the ground and water reservoirs.

Actions directed at struggle against formation and spread of dust can be divided into:
 – application of processes which result in minimal material crushing,
 – maximal mechanization and automation of production processes (it does not reduce the amount of dust, but allows to reduce to a minimum the number of workers which are in the zones with intensive dust formation),
 – utilization of hermetic equipment, hermetic devices for transportation of dust-creating material,
 – moistening of dust-creating material,
 – application of filter installations, etc.

Basic directions of dust-reduction measures in open cast collieries are methods of struggle against dust, which are based on the use of water. Efficiency of these methods is determined by the condition that with the increase of material humidity its dust-formation ability is sharply reduced.

Various superficially active substance (PAV) for example wetting agent DB sintanol DT-7, neonole etc can be added to water used for dust control. By reducing water superficial tension the efficiency of dust control is increased. If dust control by means of water is conducted in sub-zero temperatures, various salts are added to prevent water from freezing. In addition the salts reduce the speed of water evaporation and so increase the time of its moistening action. Some salts (for example, CaCl$_2$) have hygroscopic properties, namely they are capable to absorb moisture from the air, and this property is more evident in a humid atmosphere.

The most powerful source of dust control by means of water in open cast collieries are drilling works if they are done without the application of dust catching methods. However according to

Safety rules relating to extracting of mineral deposits using open cast method (PB 03-498-02), work of drilling machines without effective dust control or dust catching methods is prohibited.

In drilling machines used in open cast collieries systems using compressed air are usually applied for clearing boreholes from cuttings. The aerosol formed during drilling also requires clearing.

The manufacturer makes drilling machines complete with a dust catching system at the factory. Dust catching systems from drilling machines have three levels of clearing:

I – clearing of cuttings and big fractions of dust,

II – clearing of medium and fine fractions of dust,

III – clearing of thin fractions of dust.

For the first level dust collecting caps or dust collectors are used, catching up to 95% of cuttings (having size bigger than 500 microns), for the second level cyclones having 90% efficiency of catching dust particles with size bigger than 10 microns are applied, and for the third level fabric filters are used with efficiency of catching thin dust fractions (size smaller than 10 microns) 99% and more.

Decision as to the choice of other supplementary elements of dust catching systems and establishment of clearing levels is made depending on the ability of dust particles formed during drilling to become airborne. Efficiencies of various dust catching methods for particles of different sizes are specified in table 2.

Table 2. Areas of utilization of different dust control devices

DUST CONTROL DEVICE	SIZE OF PARTICLES [microns]					
	>3000	500–3000	60–500	5–60	0,1–5	<0,1
Dust-collecting chambers (umbrellas, bunkers)						
Cyclones						
Multicyclones and groups of cyclones of small diameter						
Fabric filters						
Water						
Ultrasonic coagulators	do not coagulate					
Electro-filters	application is not rational					

Large quantities of dust and noxious gases, which are in form of dust-and-gas clouds, are formed in mining works' atmosphere by blasting. The cloud temperature during the first instants after explosion considerably exceeds ambient temperature, and therefore it moves upwards. The cloud reaches the height of 150–250 meters in 20–30 seconds then starts to be moved by the wind. Thus intensive separation of dust particles occurs. A great quantity of dust settles at the distance of 300–500 m from the place where blasting took place i.e. does not leave the limits of the works.

Dust concentration in the cloud depends on many factors, and after 30–60 seconds after the explosion and the distance of 30–40 m from the blast area it can range from 600 to 5000 mg/m^3. The quantity of gases ejected into the atmosphere depends on the volume of blasted rock and the quantity of explosives. The released gases are characterized by high concentration of noxious substances. Gaseous products of explosive decomposition can contain water vapour, carbon dioxide

CO_2, nitrogen N_2, oxygen O_2, carbon oxide CO, nitrogen oxides NO, NO_2, N_2O_4 and N_2O_5. In addition, formation of insignificant quantities of cyanic hydrogen HCN, methane CH_4, ethane C_2H_6, acetylene C_2H_2, hydrogen H_2, sulphur dioxide SO_2, hydrogen sulphide H_2S and mercury vapour is possible.

Carbon oxide is a colourless gas, without taste and smell, lighter than air, less dissolved in water, boils at 192°C. It is capable of combining with haemoglobin in human blood. The ability of haemoglobin to bind with carbon oxide is 30 times bigger than with oxygen.

Nitrogen oxides resulting from blasting operations are in the form of different combinations of oxygen with nitrogen, and the most stable of them are NO_2 and N_2O_4. Both oxides during prolonged presence in humid atmosphere of mines change to nitrogen and nitrous acids, thus their poisonous influence weakens as a result of small concentrations. At strong concentrations, nitrogen oxides can be easily seen in the air as brown-coloured steams, and the higher is the temperature, the stronger is the concentration. Common character of nitrogen oxide influence on organisms is expressed by irritation of lungs, which causes their inflammation.

There are methods of conducting blasting works preventing (or mineralising) formation of these gases or providing their neutralization at the moment of formation which, when applied, decrease quantities of toxic gases formed during blasting. The most comprehensible methods are the following:
- application of explosives with zero oxygen balance;
- introduction of special additives (KCl, $NaCl$, KNO_3) into explosives structure;
- application of neutralizers: solid substances (mainly salts of alkaline metals) which are introduced in blast holes, or liquid substances sprayed on the bank before the explosion and creating a veil for blast gases and dust.

To decrease the quantity of dust in dust-and-gas cloud technological and water control measures are applied.

Among technological actions are: detonation of holes of smaller diameter and long length, detonation of charges with air intervals between them, reduction of the specific consumption of explosives and detonation in a heavy environment or on the ungathered mountain mass. As a result of use of these methods the rise height of dust-and-gas clouds and dust formation are reduced.

The methods utilizing water for dust suppression in mass explosions can be divided into two phases – during explosion and after explosion. The first phase refers to internal and external water tamping and water blasting, to the second phase – preliminary watering of the areas adjacent to the blasting area.

Besides that, for the intensification of dust formation processes from broken rock mass and simultaneously to facilitate spreading of dust-and-gas clouds, blasting works must coincide with maximum wind activity and utilize the application of artificial ventilation of pits (these methods do not reduce dust and gas formation). For pit ventilation it is expedient to apply installations, which create free water-air jets and provide for intensification of gas formation process with simultaneous suppression and prevention of dust pulverization.

Loading and off-loading works and transport are one of the basic permanent sources of dust formation and dust separation in coal works. During loading of broken rock mass dust elimination exceeds dust formation.

To minimize the formation of dust during cargo-handling works different types of watering and also preliminary humidifying of loaded material are used. Warm or cold water, water solutions of salts ($NaCl$, $CaCl_2$, KCl, etc.) and PAV are used for watering and preliminary humidifying. Watering is done with the help of various sprinklers, which can be positioned on cargo handling equipment (for example, on a dredge arrow) or can be set up as independent mobile or stationary watering installations. Preliminary humidifying can be done on solid rock mass by pumping liquids into specially drilled holes or after completion of blasting works by flooding the liquid on the surface of the loosened and levelled mass of rock. Intensity of dust separation during work of excavation equipment reaches: excavator – 2 g/s, excavator – dragline 11 g/s, discharging dump car 64–275 mg/s and bulldozer 97–200 mg/s, etc.

Transport of rock and coal by car is a cause for dust separation occurring as a result of abrasion and crushing, spill, small particles of rock brought onto roads by car wheels, skidding of car wheels, eddy flows formed by car movement, and also by wind erosion of the ramps. Intensity of dust separation depends on physico-mechanical properties of material from which car ways are made, weight, type, intensity and speed of car movement, parameters and road condition, wind speed and direction, other factors.

Radical method of struggle against dust on car ways is the use of roads with concrete surface. Application of such road surfaces in combination with their periodic cleaning, permits to reduce dust quantities in the air to MAC. However this method is connected to large investment expenses and it is not always applicable.

Table 3. Methods and parameters of application of dust adhesive substances on roads

CONDITIONS OF APPLICATION	RECOMMENDED SUBSTANCES FOR DUST ADHESIVENESS	METHOD OF APPLICATION	CONSUMPTION FOR 1 m^2 OF THE SURFACE [kg]	TIME OF ACTION OF ONE PROCESSING [day]
Air temperatures above zero, small or moderate quantity of rainfalls	Universin-L (summer)	Impregnation or superficial processing of loosened dry road covering with subsequent levelling	0,7–2,0	10–30
	Lignosulphonate	Superficial processing of profiled road with subsequent levelling	1,5–3,0	up to 20
	Mixture of water, lignosulphonate (5–10% mass) and polyacrylamide (0,5–0,2% mass)	Irrigation of profiled and levelled road surface with subsequent processing with 0,2–0,5% of polyacrylamide solution and levelling	6	5–10
	Lignodor	Superficial processing of profiled road with subsequent levelling	2,0–2,2	up to 45
	Bitumen emulsions EBA-3	Superficial processing, clearing from dust, plane and levelled road surface with subsequent covering with sand or with fine rubble	0,8–1,5	up to 30
	Clay	Covering the prepared layer with dry clay with subsequent periodic (1 time per shift) levelling with water	50–80	7–15
		Covering the prepared layer with suspension clay with subsequent periodic (2 times per shift) moistening with water	16–18 kg of clay (40–50 l suspense)	6–10
	Bitumen SG, MG, MGO (40/70)	Impregnation of loosened dry surface with subsequent levelling	0,2–2,0	10–30
Air temperatures above zero, large number of rainfalls	Hygroscopic salts	Irrigation of road surface	0,8–2,5	5–15
	Universin-V (high-viscid)	Impregnation of loosened dry surface with subsequent levelling or unloosened surface with subsequent covering with fine rubble and levelling	1,5–2,5	20–30
	Bitumen SG, MG, MGO (70/130)	Road construction device using the method of hashing on the road	7–8	40–60
Air temperatures far below zero, small quantities of rainfalls	Universin-S	Superficial processing of loosened (if necessary) road surface with subsequent levelling	1–2	5–15
Sub-zero air temperatures, large number of rainfalls	Hygroscopic salts	Introduction in superficial layer of carriage road with subsequent levelling and hashing	2,5–3,5	10

Reduction of dust at quarry car roads can be achieved also by the use of dust-binding substances, which can be divided into two groups: hygroscopic salts (CaCl₂ etc.) and organic dust-binding substances. Characteristics of dust adhesive substances are represented in table 3.

In sub-zero temperatures the most effective, ecologically safe and technological method of dust control is strengthening surfaces of ramps with cold water.

An additional source of dust formation is pulverization of settled dust by the wind blowing it from the surfaces of broken rock, ledges, landslides and also the rock, which is in cars or in railway cars. The intensity of dust pulverization process depends on different factors: degree of disperssiveness of the motes and their form, mineralogical and chemical composition of dust, specific weight, force of binding with the surface and speeds of air stream. However the determining factor is air stream speed (fig. 1).

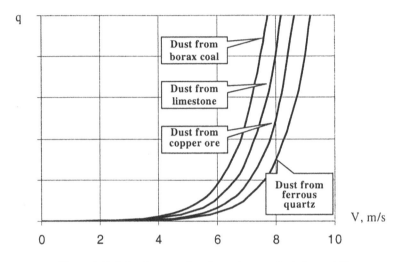

Figure 1. Relation between dust formation q and wind speed V

To prevent dust from being blown away, dust particles need to be treated with reagents (latex, liquid glass, clay, hygroscopic salts, etc.). By such processing a film is formed, which prevents dust from being blown. Also the same methods of dust control as used on roads can be applied.

Some machinery and vehicles used in mining and in coal works are equipped with diesel engines, which are sources of emission of toxic exhaust gases (EG) into the environment. It is known that there are more than 200 components determining the qualitative difference of EG from atmospheric air, but toxic mixtures make up only about 1% of the total weight of EG. In the biggest works, however, the emissions of a big number of simultaneously working diesel engines can make the atmosphere unsuitable for breathing. In deep pits, emissions of EG from diesel engines is one of the factors limiting application of machinery equipped with diesel engines at the bottom horizons. EG toxic components are CO, nitrogen oxide, aldehydes, hydrocarbons, sulfur oxide, soot, and 3, 4 – benzo(a)pirene.

Hence, coal mining by open cast method is a powerful source of pollution to the atmosphere that causes the necessity to undertake actions at coal works aiming at atmosphere protection. Table 4 presents the average structure of basic components of gas-elimination and their toxicity graduation.

Table 4. Average structure of basic components of EG of diesel engines

COMPONENT	COMPOSITION		TOXICITY
	% on volume	g/m^3	
Nitrogen, N$_2$	76–78	950–975	Nontoxic
Oxygen, O$_2$	2–18	30–260	Nontoxic
Water vapour, H$_2$O	0,5–4,0	4–32	Nontoxic
Carbon dioxide, CO$_2$	1–10	20–200	Nontoxic
Hydrogen, H$_2$	0–0,05	0–0,04	Nontoxic
Carbon oxide, CO	0,01–0,5	0,12–6,25	Toxic
Nitrogen oxides (recounting on NO$_2$)	0,001–0,4	0,02–1,0	Toxic
Aldehydes (recounting formaldehyde H$_2$C = CH-CH = O$_2$)	0–0,002	0–0,03	Toxic
Hydrocarbons, C$_x$H$_y$ (total)	0,002–0,02		Toxic
Sulphur dioxide, SO$_2$	0–0,003	0,86	Toxic
Carbon, C		0,01–1,5	Toxic
3, 4-benzo(a)pirene, C$_{20}$H$_{12}$		up to 10^{-5}	Toxic

By the character of impact on the environment, people, chemical structure and properties, EG components are divided into groups.

I group – nontoxic substances: nitrogen N$_2$, oxygen O$_2$, hydrogen H$_2$, water vapour H$_2$O, carbon dioxide CO$_2$ (harmful because it is one of the gases which create greenhouse effect).

II group – carbon oxide CO. It influences human organism depending on its concentration in air.

III group – nitrogen oxides, mainly oxide and dioxide. Nitrogen oxide NO – colourless gas. Nitrogen dioxide NO$_2$ – brown coloured, with characteristic smell, heavier than air, boils at 21°C. Nitrogen oxides participate in photochemical reaction of smog formation. They influence human organism depending on their volumetric concentration in the atmosphere.

IV group – hydrocarbons C$_x$H$_y$, being representatives of all homological types: alkanes, alkenes, alkadienes, cyalkanes, aromatic combinations, including carcinogens. Hydrocarbons are toxic and participate in photochemical reactions with nitrogen oxides. Among multinuclear aromatic hydrocarbons (MAH) with condensed rings the most widespread is 3,4 benzo(a)pirene C$_{20}$H$_{12}$, characterized by high cancerogenic activity.

V group – aldehydes. The ones present in EG are basically formaldehyde and acrolein. Formaldehyde is a colourless gas with a pungent smell, specific weight in relation to air – 1,04, easily dissolved in water. Acrolein (or acryl acid aldehyde) – is a colourless liquid having the smell of burnt grease, boiling temperature 52,4°C, easily dissolved in water. It influences human organism depending on its concentration in the atmosphere.

VI group – carbon, formed during burning as a result of pyrolysis (thermal disintegration) of hydro-carbonic molecules in the condition of lack of oxygen. Carbon in its pure state does not represent any danger to humans, however it is a carrier of carcinogenic hydrocarbons, because it absorbs them very well due to its developed surface.

3,4-benzo(a)pirene, adsorbed by the surface of carbon, influences live cells stronger than in its pure state.

Into the composition of diesel EG when using sulphurous carburant also enter inorganic gases: sulphur dioxide SO$_2$ and hydrogen sulphide H$_2$S.

Thus, the toxicity and irritating properties of diesel EG, taking into account quantitative structure of micro-mixtures, is practically completely defined by five harmful components: nitrogen dioxide NO$_2$, carbon oxide CO, hydrocarbons C$_x$H$_y$, sulphur dioxide SO$_2$ and carbon C.

Efforts aimed at decreasing diesel harmful influence on the environment can be divided into two basic directions, which suppose:
- prevention of formation of toxic substances during combustion of fuel in diesel cylinders,
- external processing of EG before its release in the atmosphere.

Other methods of external processing of EG are:
- neutralization as a result of which toxic EG components are converted in non-toxic,
- clearing as a result of which harmful components are caught by mechanical, electric or other filters,
- diluting EG with atmospheric air, which is seldom applied as an independent method, because it does not reduce the quantity of harmful emissions into the atmosphere and it is used as addition to the first two methods, increasing their efficiency, decreasing the concentration of EG components and their temperature to background values.

The most widespread methods of cleaning of EG car diesel engines are catalytic and liquid neutralization. Wide application of these methods is limited however by the following factors: catalytic neutralizers use expensive platinum-nickel catalysts; liquid neutralizers for EG clearing use water, which becomes very dirty and before it is dumped or reused requires complex and expensive treating. The quantity of water for liquid neutralization is established by the capacity of transport car engines.

In this way, coal mining using open cast method is a powerful source of atmosphere pollution with dust and toxic gases. Pollution sources of the atmosphere are all technological processes done in open cast mining works. Various solutions for protection of the atmosphere can be applied depending on the features of dust-and-gas formation at these processes.

REFERENCES

Slastunov S.V., Koroleva V.N., Kolikov K.S., Kulikova E.J., Vorobiev A.E., etc.: Mining and Environment. Textbook. M.: Logos 2001.

Feldman J.G.: Hygienic Estimation of Auto Transport, as the Source of Atmospheric Air Pollution. M.: Medicine 1985.

Guide. Open Mountain Works. Trubetskoj K.N., Potapov M.G., Vinitskij K.E., Melnikov N.N., etc. M.: Mountain Bureau 1994. Page 590.

Kudrjashov V.V., Umantsev R.F., Shurinova M.K.: Thermo-Humid Dust Processing of Many Years Frozen and Destroyed Coal Massive. M.: Rotaprint IPCON AN USSR 1991. Page 136.

Chulakov P.C.: Theory and Practice of Dust Careers Atmospheres. M.: Bowels 1973. Page 160.

Ishuk I.G., Pozdnyakov G.A.: Methods of Complex Dust Elimination from Mining Enterprises: Guide. M.: Nedra 1991. Page 253.

Ivashkin V.S. Struggle Against Dust and Gases from Coal Open Cast Collieries. M.: Nedra 1980. Page 152.

Filatov S.S., Mihajlov V.A., Vershinin A.A.: Struggle Against Dust and Gases From Quarry. M.: Nedra 1973. Page 144.

Loboda I.A., Rebristyi B.N., Tyschuk V.J. etc.: Struggle Against Dust in Open Cast Mining Works. Kiev. Tehnika 1989. Page 152.

Ushakov K.Z., Kaledina N.O., Kirin B.F., Srebnyi M.A.: Safety of Vital Activity. Book for High Schools. Under K.Z Ushacov. M.: Printing House MGGU 2000. Page 430.

Safety Rules for Development of Useful Mineral Deposits by Open Cast Method. PB 03-498-02.

Osodoev M.T.: Struggle Against Dust in Yakutia Coal Open Cast Collieries. Yakutsk: Printing House IF SO AN URSS 1987. Page 116.

Author Index

Printed and bound by CPI Group (UK) Ltd, Croydon, CR0 4YY

01/11/2024

01782599-0016